ADVANCE PRAISE FOR
Choose Economic Freedom

"George Shultz and John Taylor's parable reminds us of a time when Americans suffered greatly from bad economic policy choices. Most people, honorable men and women included, had convinced themselves that those policies were called for. It turned out that most people were wrong. I was there for this story—but I learned much from this book that is new and highly relevant to today, and so will you."

—**Alan Greenspan**, former chair, Council of Economic Advisers and Federal Reserve Board

"John Taylor is a damn good economist steeped in public policy. His name alone would guarantee a world-class rendition of US policies in the 1970s and 1980s. But when John Taylor is combined with George Shultz—my candidate as the single most important public figure in the last one hundred years—you not only have the truth, but you have the whole truth.

The Nixon years and the Reagan years stand in stark contrast when it comes to policies and outcomes. Good policy is not about party or ideology; it's all about economics, and economics is all about incentives. This book debunks the notion that facts don't matter. Facts do matter, and as this book shows, the story facts tell is as clear as clear can be. Markets work.

This book by two of the most honest and knowledgeable policy experts gives you much more than just a peek behind the opaque curtains of the US 'C' suite. It's a full-blown exposé."

—**Arthur B. Laffer**, former chief economist at the Office of Management and Budget and member of President Ronald Reagan's Economic Policy Advisory Board

"John Taylor and George Shultz have done a great service in *Choose Economic Freedom*. The book covers a lot of territory—my whole economic career—and a lot of interesting people. The writing is clear and, yes, fun to read. What you will not get from them are 'guideposts,' 'wage and price guidelines,' or offers to get something valuable without working for it. Here we are with puritans like Milton Friedman and George Shultz, University of Chicago colleagues from the 1960s. We read of Shultz resigning his position under Nixon when Nixon continued to support what he (and Friedman) thought to be bad economics. There followed ten years of pointless inflation with Nixon, Ford, and Carter. Then, finally in 1980, Ronald Reagan, Paul Volcker, and Shultz wound the inflation down back to what it could have been in 1970. What a waste, but at least that's over. Well . . . maybe you had better first look at Taylor's analysis of 1990 or 2000!"

—**Robert E. Lucas Jr.**, John Dewey Distinguished Service Professor Emeritus in Economics and the College at the University of Chicago, and 1995 recipient of the Nobel Memorial Prize in Economic Sciences

Choose Economic Freedom

Choose Economic Freedom

Enduring Policy Lessons from the 1970s and 1980s

GEORGE P. SHULTZ
AND JOHN B. TAYLOR

with Words of Wisdom by Milton Friedman

HOOVER INSTITUTION PRESS
STANFORD UNIVERSITY   STANFORD, CALIFORNIA

With its eminent scholars and world-renowned library and archives, the Hoover Institution seeks to improve the human condition by advancing ideas that promote economic opportunity and prosperity, while securing and safeguarding peace for America and all mankind. The views expressed in its publications are entirely those of the authors and do not necessarily reflect the views of the staff, officers, or Board of Overseers of the Hoover Institution.

hoover.org

Hoover Institution Press Publication No. 708

Hoover Institution at Leland Stanford Junior University,
Stanford, California 94305-6003

First printing 2020
26 25 24 23 22 21 20   7 6 5 4 3 2 1

Manufactured in the United States of America

The paper used in this publication meets the minimum Requirements of the American National Standard for Information Sciences—Permanence of Paper for Printed Library Materials, ANSI/NISO Z39.48-1992. ♾

Library of Congress Control Number:2019956973

ISBN: 978-0-8179-2344-0 (cloth. : alk. paper)
ISBN: 978-0-8179-2346-4 (epub)
ISBN: 978-0-8179-2347-1 (mobi)
ISBN: 978-0-8179-2348-8 (PDF)

CONTENTS

Preface   vii

CHAPTER ONE
Guidelines: Debated and Implemented   1

CHAPTER TWO
Wage and Price Controls: Bad Economics Brings
Bad Policy   17

CHAPTER THREE
Damage Control as the Economy Fights Back   27

CHAPTER FOUR
Reversal and Reform: Good Economics Generates
Good Policy   35

CHAPTER FIVE
Orthodox Policies and Strategic Thinking Work   55

APPENDIX A

Timeline of Key Events 65

APPENDIX B

Letter from Arthur Burns to President Nixon,
June 22, 1971 73

APPENDIX C

Economic Strategy for the Reagan Administration:
A Report to President-Elect Ronald Reagan from His
Coordinating Committee on Economic Policy,
November 16, 1980 81

APPENDIX D

Frequently Asked Questions 95

Notes 103
References for Further Reading 105
Acknowledgments 107
About the Authors 109
Index 113

PREFACE

THIS BOOK examines essential lessons learned from economic policies that were tried and that failed in the 1970s and reversed with success in the 1980s. We focus on government interventions into wage and price setting in the private economy—guideposts, freezes, controls, and so on—but we show that the same lessons apply to other government interventions and in other time periods, including the present.

The book starts in chapter 1 with a debate between economists Milton Friedman and Robert Solow in 1962 about the use of wage and price guideposts as a tool of government policy. Those 1960s guideposts were the precursor of outright price and wage freezes and controls in the 1970s. Friedman argued against the wage and price guideposts and Solow argued in favor of them. Verbatim quotes from the debate show how two brilliant economists can be diametrically opposed about a major policy because of differences of view in how the economy works. Eventually, the debate led to major changes in the way economists model the overall economy, largely as Friedman suggested—although, as we show in later retrospectives by the two economists, the disagreements did not go away.

Chapter 2 shows how the guidepost policy of the 1960s expanded into wage and price control policy in the 1970s, starting with a freeze in 1971. Again, views about how the economy worked featured prominently in the policy decision. In a 1971 memo newly unearthed in the Hoover Archives (included as appendix B), Arthur Burns, economic adviser to President Nixon and chair of the Federal Reserve Board, argued in a letter to the president, marked "personal and confidential," that the private economy had changed: it now needed government interventions into private-sector wage and price decisions if it were to experience low inflation and low unemployment. Burns did not think the Fed's traditional monetary policy tools could do the job, so he advocated for government controls on wages and prices—and he won the argument. Though many then thought of Burns as being essentially infallible, the chapter opening, which describes a White House engagement with George Shultz, shows that he was not so infallible after all.

Chapter 3 delves into the costs of the wage and price controls as enacted in practice. Although the controls were popular at first, the Federal Reserve under Burns began to increase money growth, and inflation rose. So controls were reimposed in 1973. By this time, a complex bureaucratic apparatus had been created to monitor and administer prices, and the system had become a drag on the economy, as Friedman had predicted. Shultz resigned as secretary of the Treasury in 1974 in protest, but the damage continued into the Ford administration with more failed and unorthodox inflation-control policies—such as the "Whip Inflation

Now" buttons—and on through the rest of the 1970s as described by John Taylor, who worked at the Council of Economic Advisers during this period.

Chapter 4 describes the reversal of the policy and the end of the controls. In a 1974 speech, Friedman contrasts the costs of the interventionist wage and price controls with the benefits of the noninterventionist move toward flexible exchange rates that he showed was due to Shultz's leadership as secretary of the Treasury. How much better would the economy have been, Friedman asked, if that market-oriented approach had been followed in the domestic economy?

But it was not until the start of the Reagan administration that all the controls were removed, including those on energy, and the guideposts were ended. The end of price controls and the support for sound monetary policy were central elements of a set of recommendations contained in a memo (included as appendix C) to President-elect Reagan from economists who had advised him in the campaign. By this time, Paul Volcker was chair of the Federal Reserve Board. He began to introduce a sound monetary policy to reduce inflation, essentially reversing the inflationary trajectory started under Burns. There was no need for interventionist wage and price controls when these tried-and-true methods were in place. The chapter ends as it began, with a crystal-clear analysis from Friedman expressed in a later interview. He contrasts the approaches used under President Nixon in the 1970s, when Burns was Fed chair, with the policy approach under President Reagan in the 1980s, with Volcker as Fed chair.

Chapter 5 draws the main lesson from this important era in US history. The evidence shows that tried-and-true economic policy works and that, however tempting at any given time, deviating from such policy doesn't. The starting point for deviation is often a new view of how the economy works, which calls for a different policy approach, but which proves wrong too late, with lasting damage. The needed policy reversal is often difficult, because views get entrenched, but the eventual improved performance is usually convincing.

In our experience in government, Shultz's mainly in the 1970s and 1980s and Taylor's mainly in the 1990s and 2000s, this dynamic sequence is common in all areas of economic policy making, not simply wage and price controls. The breadth is reflected in the comprehensive set of reforms—tax, regulatory, budget, monetary—listed in the 1980s policy memo in appendix C. Moreover, in our experience, the policy cycle is recurring: similar stories are found in recent years, from the 1990s to the present. It requires vigilance on the part of policy makers and an insistence on a robust, open analytical approach in order to avoid costly deviations from good economic policy making. In other words, let's examine this little piece of economic history and see what it teaches us about future economic policy.

CHAPTER ONE

Guidelines

Debated and Implemented

L ET'S BEGIN in 1962. Those were the days of a new and energetic Kennedy administration wanting to expand the economy but worried about inflation. In 1962, the president's Council of Economic Advisers used their annual *Economic Report of the President*[1] to issue a set of guidelines for wage and price changes designed to set the sights of labor and management on wage and price changes in a way that would keep inflation under control. They outlined voluntary, productivity-linked "guideposts" to stabilize wage and price setting in industries they said exercised market power. The concerns about inflation were underlined by Lyndon Johnson's Vietnam War and Great Society spending. People called it a guns-and-butter economy and continued to worry about inflation. The White House increasingly found itself involved in manufacturing-industry wage negotiations and in publicly warning businessmen not to raise consumer prices. Concerned that these guidelines might lead to wage and price controls, George Shultz and his Chicago colleague Robert Aliber pulled together a conference on the subject in April 1966. It turned out to be a very productive meeting attended by top-notch economists of all persuasions, including George Stigler, Allan Meltzer, then CEA chair Gardner Ackley, and future labor secretary John Dunlop. The papers and proceedings were published in a book, *Guidelines, Informal Controls, and the Market Place.*[2] The discussions were intense but not reassuring. There was widespread support for the guidelines, little concern that they might lead to wage and price controls (see appendix D), and an almost "So what?" attitude toward such controls if they materialized.

Many influential people attended the conference, but the stage was set by Milton Friedman, then of the University of Chicago, who would win the Nobel Prize in economics in 1976; and Robert Solow of the Massachusetts Institute of Technology, who would win the Nobel Prize in 1987. We quote these economists at length because these two brilliant thinkers broadly argue the merits of intervention and their extended statements are worthy of consideration. The complete document of their speeches and comments on each other's statements is found in the book *Guidelines, Informal Controls, and the Market Place.*

In his statement, Friedman argues against wage and price guidelines or guideposts, pointing to the harms they cause— including that it is in the business's and the economy's interest to violate government requests (and perhaps even legal mandates)—and showing that inflation is a "monetary phenomenon" caused by bad monetary policy:

> Entirely aside from their strictly economic effects, guidelines threaten the consensus of shared values that is the moral basis of a free society. Compliance with them is urged in the name of

social responsibility; yet, those who comply hurt both themselves and the community. Morally questionable behavior—the evading of requests from the highest officials, let alone the violation of legally imposed price and wage controls—is both privately and socially beneficial. That way lies disrespect for the law on the part of the public and pressure to use extralegal powers on the part of officials. The price of guideposts is far too high for the return, which, at most, is the appearance of doing something about a real problem....[3]

Yet, the central fact is that inflation is always and everywhere a monetary phenomenon. Historically, substantial changes in prices have always occurred together with substantial changes in the quantity of money relative to output. I know of no exception to this generalization, no occasion in the United States or elsewhere when prices have risen substantially without a substantial rise in the quantity of money relative to output or when the quantity of money has risen substantially relative to output without a substantial rise in prices. And there are numerous confirming examples. Indeed, I doubt that there is any other empirical generalization in economics for which there is as much organized evidence covering so wide a range of space and time.[4]

Friedman then compares the harms of "open" inflation (without wage and price guidelines) with the harms of "suppressed" inflation (with wage and price guidelines) noting that guidelines make the problem worse:

Open inflation is harmful. It generally produces undesirable transfers of income and wealth, weakens the social fabric, and may distort the pattern of output. But if moderate, and especially if steady, it tends to become anticipated and its worst effects on the distribution of income are offset. It still does harm, but, *so long as prices are free to move,* the extremely flexible private enterprise system will adapt to it, take it in stride, and continue to operate efficiently. The main dangers from open inflation are twofold: first, the temptation to step up the rate of inflation as the economy adapts itself; second, and even more serious, the temptation to attempt cures, especially suppression, that are worse than the disease.

Suppressed inflation is a very different thing. Even a moderate inflation, if effectively suppressed over a wide range, can do untold damage to the economic system, require widespread government intervention into the details of economic activity, destroy a free enterprise system, and along with it, political freedom. The reason is that suppression prevents the price system from working. The government is driven to try to provide a substitute that is extremely inefficient. The usual outcome, pending a complete monetary reform, is an uneasy compromise between official tolerance of evasion of price controls and a collectivist economy. The greater the ingenuity of private individuals in evading the price controls and the greater the tolerance of officials in blinking at such evasions, the less the harm that is done; the more law-abiding the citizens, and the more rigid and effective the governmental enforcement machinery, the greater the harm. . . .[5]

The guideposts do harm even when only lip service is paid to them, and the more extensive the compliance, the greater the harm.

In the first place, the guideposts confuse the issue and make correct policy less likely. If there is inflation or inflationary pressure, the governmental monetary (or, some would say, fiscal) authorities are responsible. It is they who must take corrective measures if the inflation is to be stopped. Naturally, the authorities want to shift the blame, so they castigate the rapacious businessman and the selfish labor leader. By approving guidelines, the businessman and the labor leader implicitly whitewash the government for its role and plead guilty to the charge. They thereby encourage the government to postpone taking the corrective measures that alone can succeed.

In the second place, whatever measure of actual compliance there is introduces just that much distortion into the allocation of resources and the distribution of output. To whatever extent the price system is displaced, some other system of organizing resources and rationing output must be adopted. As in the example of the controls on foreign loans by banks, one adverse effect is to foster private collusive arrangements, so that a measure undertaken to keep prices down leads to government support and encouragement of private monopolistic arrangements.

In the third place, "voluntary" controls invite the use of extra-legal powers to produce compliance [see appendix D]. And, in the modern world, such powers are ample. There is hardly a business concern that could not have great costs imposed on it by antitrust investigations, tax inquiries, government boycott, or rigid enforcement of any of a myriad of laws, or on the other side of the ledger, that can see no potential benefits from government orders, guarantees of loans, or similar measures. Which of us as an individual could not be, at the very least, seriously inconvenienced by investigation of his income tax returns, no matter how faithfully and carefully prepared, or by the enforcement to the letter of laws we may not even know about? This threat casts a shadow well beyond any particular instance. In a dissenting opinion in a recent court case involving a "stand-in" in a public library, Justice Black wrote, "It should be remembered that if one group can take over libraries for one cause, other groups will assert the right to do it for causes which, while wholly legal, may not be so appealing to this court." Precisely the same point applies here. If legal powers granted for other purposes can today be used for the "good" purpose of holding down prices, tomorrow they can be used for other purposes that will seem equally "good" to the men in power—such as simply keeping themselves in power. It is notable how sharp has been the decline in the number of businessmen willing to be quoted by name when they make adverse comments on government.

In the fourth place, compliance with voluntary controls imposes a severe conflict of responsibilities on businessmen and labor leaders. The corporate official is an agent of his stockholders; the labor leader, of the members of his union. He has a responsibility to promote their interests. He is now told that he must sacrifice their interests to some supposedly higher social responsibility. Even supposing that he can know what "social responsibility" demands—say by simply accepting on that question the gospel according to the Council of Economic Advisers—to what extent is it proper for him to do so? If he is to become a civil servant in fact, will he long remain an employee of the stockholders or an agent of the workers in name? Will they not discharge him? Or, alternatively, will not the government exert authority over him in name as in fact? . . .[6]

Milton Friedman receives the Nobel Memorial Prize in Economic Sciences in December 1976. *Jan Collsioo/TT/Sipa USA. Milton Friedman papers, Box 113, Folder 18, Hoover Institution Archives.*

Guideposts and pleas for voluntary compliance are a halfway house whose only merit is that they can more readily be abandoned than legally imposed controls. They are not an alternative to other effective measures to stem inflation, but at most a smokescreen to conceal the lack of action. Even if not complied with they do harm, and the more faithfully they are complied with, the more harm they do.[7]

In response, Bob Solow makes the case *for* guidelines—or, as he put it, "the case against the case against" guidelines made by Friedman. Guidelines, which he sometimes calls "incomes policy," are needed, he argues, because modern economies are more mixed—they are less competitive than Friedman argues, so inflation would be too high when the economy operates at full employment:

> The problem is that modern mixed capitalist economies tend to generate unacceptably fast increases in money [i.e., nominal] wages and prices while there is not general excess demand. No particular view of the economic process or of the determinants of demand need be implied by this observation. It is a fact, however, or at least it is widely believed to be a fact, that wages and prices begin to rise too rapidly for comfort while there is still quite a bit of unemployed labor and idle productive capacity and no important bottlenecks. This tendency creates a dilemma for public policy. Governments generally do not wish to acquiesce in an inflationary spiral; indeed, in our rather international trading world, governments may not be able to do so. On the other hand, governments value employment and output, for the very good reason that people value employment and output, so governments generally do not wish to choke off economic expansion while there is room for more. . . .[8]

It is not an accident that the Council of Economic Advisers launched the local version of incomes policy in the January 1962 Economic Report, despite the fact that the unemployment rate was then near 6 percent and manufacturing capacity only 83 percent utilized, according to the McGraw-Hill survey.[9] Wholesale prices were not then rising, nor did they begin to rise until 1965. But still the Council felt—with good reason—that it had to protect its flank against those who argued, even then, that an expansionary fiscal and monetary policy would dissipate itself almost immediately in inflationary wage and price behavior. The argument proved wrong; but that it could be seriously made suggests the nature and the seriousness of the problem of premature inflation. . . .[10]

If there is a case against the case against the guideposts, part of it has to be that the first obvious remedy may be very costly, and the second obvious remedy is more than a little unrealistic.

The experience of the years 1958–64 certainly indicates that the economy can be run with quite a lot of slack, but not a catastrophic amount, so that the price level will more or less police itself. That is a possible policy. But it is not a costless policy. . . . This is a budgetary cost, but not a real burden on the economy as a whole. In the second place, however, the maintenance of slack does represent a real burden to the economy as a whole in the form of unproduced output. It is not easy to make any estimate of that cost. The usual rule of thumb is that one-half point on the unemployment rate corresponds to something between 1 and 2 percent of real GNP. In that case, the amount of relief from inflation that could be had by keeping the unemployment rate one-half point higher than otherwise desirable would have an annual cost of about $10 billion at 1965 prices and GNP. . . . But it does mean that an alternative policy capable of having the same restraining effect as a half point of unemployment is a preferable policy unless it imposes social costs of about that order of magnitude.[11]

Next Solow endeavors to explain why he thinks firms with market power—that is, with few competitors—will drive up prices and cause inflation even when there is slack in the economy:

> The logic of a guidepost policy is, I suppose, something like this. In our imperfect world, there are important areas where market power is sufficiently concentrated that price and wage decisions are made with a significant amount of discretion. When times are reasonably good, that discretion may be exercised in ways that contribute to premature inflation. (Institutions with market power may actually succeed in exploiting the rest of the economy temporarily or permanently, or they may see their decisions cancelled out almost immediately by induced increases in other prices and wages.) People and institutions with market power may, in our culture, be fairly sensitive to public opinion. To the extent that they are, an educated and mobilized public opinion may exert some restraining pressure to forestall or limit premature inflation. . . .[12]

> The object of the guideposts was and is to hold up to the public—and to those participants in wage and price decisions who can exercise some discretion—a summary picture of how wages and prices would behave in a fairly smoothly functioning competitive market economy subject neither to major excess demand nor major deficiency of demand. The hope was that active discussion of the issues might induce the participants, in effect, to imitate a little more closely a few aspects of competitive price and wage behavior. If that happened, the expansion of real demand and the production of real output might be able to go a little further before unacceptable increases in the general price level would begin. . . .[13]

The most common criticism of dependence on wage-price guide-posts is that they simply do not work and have no effect on either wages or prices. Some of these criticisms simply cancel one another: For every employer who complains that unions take the guidepost figure for a floor, there is a union leader who complains that employers take it for a ceiling. Such evidence is worth nothing. Better evidence can be had, but is in the nature of the case uncertain. We may not be able to tell whether the guideposts have had any influence on wage and price decisions: first because there is no way to measure the "intensity" with which the guideposts have been pressed; and second because we have no universally accepted quantitative doctrine about how prices and money wages are determined in the absence of guideposts. . . .[14]

The object of the guideposts is to stall off premature inflation. Wages themselves are a matter of concern only because they bulk so large in total costs. If the guideposts served only to damp the increase in wages without holding down the price level, then their main result would simply be a transfer of income from wages to profits, and that is not their purpose. So the question arises whether there has been any visible change in price behavior.[15]

Even apart from this question of distribution, one hears it said that the guideposts are a dangerous interference in the free market, even a form of price control. At least this criticism is inconsistent with the other one that claims the guideposts to be ineffective. With some ingenuity, one could probably cook up a set of assumptions under which the guideposts had no effect on wage-price behavior yet managed to do harm to the market economy.

But this seems farfetched to me. If they are a real interference with the market, they must be partially effective.[16]

Solow then argues that he does not think guideposts are like wage and price controls. Rather, they are a way to educate the public that there is a tendency toward inflation and that firms can avoid this. He argues that the case in the 1962 *Economic Report of the President* was based on this educational argument, and that that was just fine:

> I would contend that it is also farfetched to describe the wage-price guideposts as anything remotely like a system of wage and price controls. But in any case, I am not concerned with the way the guideposts have been used by this President or that President, but with the way they were intended. They were intended, as I mentioned earlier, as a device for the education and mobilization of public opinion.
>
> That question needs to be decided on its merits. Yet, the guideposts are intended to give a summary description of a well-functioning market economy; within limits they can be expected to represent the public interest fairly well. But it is much more important to realize that the public interest does need representation. . . .[17]

This inevitable unevenness in operation strikes me as the main weakness in the guideposts. Public opinion is bound to have its greatest impact on markets that are centralized and conspicuous. That may not be all bad; centralization and discretionary power over prices and wages may be correlated. But there are obvious instances in which the correlation is broken, in which considerable market power in local markets goes along with decentralization and near-immunity to pressure from public opinion. The construction industry and the building trades unions are the standard illustration; parts of trade and transportation may provide other examples. . . .[18]

There is a different respect in which the involvement of Congress might be a good idea. Up to now, the burden of informing and mobilizing public opinion has fallen to the President and to the Chairman of the Council of Economic Advisers. This seems to be a mistake. The prestige of the President is probably too important a commodity to be spent in a way that invites occasional rebuff. And the prestige of the Council of Economic Advisers, taken by itself, is probably insufficient to carry the load. It might be helpful, therefore, if individual senators and congressmen would take part in the public debate, in their capacity as leaders and formers of public opinion. Even hearings are a possibility, provided they are hearings devoted to ordinary pieces of legislation—past or future—or to expert testimony and not to individual wage bargains or price decisions. . . .[19]

There is, as Milton said, a good deal of agreement between us. For instance, I think we both agree that when there is genuine and widespread excess demand in the economy as a whole, guidelines are almost certain to be ineffective and, if they were effective they might in fact not have entirely desirable effects.

Milton and I both agree, I think, that if we're talking about a mild inflation, it may well be both easier and better to have it open rather than repressed, provided—I would say, although Milton didn't say—that the weak, who can't protect their pensions and

Economist Robert Solow at his Nobel Prize ceremony in Stockholm, December 1987. *roger tillberg / Alamy Stock Photo.*

their savings against inflation, are in some way protected and provided for. Milton did say that the international financial policies
that the country is pursuing must be capable of withstanding that
degree of inflation.[20]

This exchange between Friedman and Solow not only set the stage for two diametrically opposed approaches to government intervention that would subsequently be put into action, it also led to a fundamental change in the way economists thought about inflation and the role of government. In an interview with John Taylor in 2000, Friedman discusses his famous 1967 presidential address before the American Economic Association, delivered the following year, in which he laid out a theory of inflation showing there is no trade-off between inflation and unemployment, as Solow had argued:

The basic ideas in my presidential address were already present in a comment that I made at a conference on guidelines, the proceedings of which were published in a 1966 book edited by George Shultz and Robert Aliber. I'm sure the basic idea grew out of the discussions about guidelines and, in particular, out of the Samuelson and Solow paper on the Phillips curve.[21] [See appendix D.] I can't say exactly where my ideas originated; all I know is by the time I gave the presidential address in 1967, there was nothing new in that compared to what I had earlier published. Arthur Burns was in the chair when I gave the presidential address, and he had gone over the address earlier. Arthur always went over my papers. . . . Despite what I said about his chairmanship of the Fed, Arthur was a first-rate economist. He had a feeling for the English language and an ability to use it, which was unusual. He was always one of the most valuable critics of anything I wrote. He didn't always agree with what I wrote, don't misunderstand me, and I'm not sure on this occasion that he agreed with me, but he was one of the people who had commented on early drafts of the paper. At the time, I never had any expectation that it would have the impact it did. It only had that impact because of the accidental factor that you had a test right after.

The "test," of course, was that the wage and price guidelines and then controls were actually put into action—the theory was tested—and the results were terrible. Inflation and unemployment both rose, as Friedman predicted, when the Fed increased money growth. Friedman's theory eventually won over the view of most economists and helped change monetary policy for the better in the 1980s.

Solow still objected, however, saying in a later debate with Taylor at the Boston Fed that Friedman had a much different assumption about the direction of causality, and Solow completely disagreed:

What Milton did without ringing any bells to warn you, was simply to take it that the causality ran the other way, that it's the deviation of the rate of inflation from the expected rate of inflation that pushes the unemployment rate away from the "natural" rate.... That kind of causality puzzled me, when I realized what had been done. Why do I say it "puzzled" me? What I mean is that I don't believe a word of it, and I found it strange that anyone does.[22]

CHAPTER TWO

Wage and Price Controls

Bad Economics Brings Bad Policy

I N 1970, Penn Central, a large financial holding and transportation company that had mismanaged its affairs, was about to file for bankruptcy. Arthur Burns, who was chairman of the Federal Reserve Board, had been chairman of President Eisenhower's Council of Economic Advisers. Burns was widely regarded as both a talented economist and an effective Washington operator. Helmut Schmidt, who would later become the chancellor of West Germany, was said to regard Arthur as the "pope of economics"—in other words, infallible.

Burns thought that if Penn Central went bankrupt there would be a great shock to the financial system. His unorthodox solution to this, which he had somehow arranged with a reluctant David Packard at the Pentagon, was what amounted to a bailout for Penn Central using Department of Defense–guaranteed V-loans. These were intended to be used by military contractors, but the company had argued it should qualify given the national security dimensions of its railroad freight operations. It was a bad idea, as it would undermine the sense of accountability for one's actions in the financial system. At a critical moment in a

White House discussion of the bailout, George Shultz, who was director of the Office of Management and Budget, recalls that into the room came President Nixon's savvy political adviser Bryce Harlow, who said, "Mr. President, in its infinite wisdom, the Penn Central Company has just hired your old law firm to represent them in this matter. Under the circumstances, you can't touch this with a ten-foot pole." So there was no bailout. The Penn Central failed . . . and guess what? The financial system was strengthened because others saw they had better get their houses in order. So the deep virtue of accountability was kept alive. Reflecting on all this, you could see that Schmidt was wrong: as impressive as he was, Arthur Burns was not infallible.

But people in Washington were talking more about inflation being intractable, and the wage and price controls were being suggested as an answer. You could sense that wage and price controls were in the air. So a speech by Shultz made the case that the budget was under control and, with a reasonable monetary policy, inflation would be brought under control. All one needed was the patience to see these policies through, so the title of the speech was "Steady As You Go." The following passage summarizes the speech:

> A portion of the battle against inflation is now over; time and the guts to take the time, not additional medicine, are required for the sickness to disappear. We should now follow a noninflationary path back to full employment.[23]

In August 1971, President Nixon announced a sixty-day wage-price freeze that was to be followed by wage and price controls.

Office of Management and Budget director George Shultz and Fed chair Arthur Burns wrestle over whether to "do something" about inflation on the cover of *Time* magazine, August 16, 1971, the week phase I wage and price controls were enacted. *From TIME. © 1971 TIME USA LLC. All rights reserved. Used under license.*

So the guideposts of the 1960s, which Friedman and Solow debated, became wage and price controls in the 1970s. This development can be understood better now than it was then: in 2018, research in the Hoover Archives revealed a letter from Arthur Burns, writing as chairman of the Federal Reserve at the time, to President Nixon. In his letter, dated June 22, 1971, Burns argued that structural changes in the economy made it difficult to control inflation, that sound monetary and fiscal policies—classical policies—would not work as in the past, and that a new approach was needed. He advocated a six-month wage and price freeze. In this private letter to the president, Burns was advocating wage and price controls. Obviously, he thought they would work, giving the Fed a major assist in taming inflation. He was an expert in the business cycle, with many years conducting research at the National Bureau of Economic Research, and this was a surprising change in his thinking. This letter, which has never before been published, is included in this volume as appendix B. Here are some key paragraphs:

1. In my judgment, some of us are continuing to interpret the economic world on the model of the 1940s and 1950s. In fact, the structure of the economy has changed profoundly since then.

   There was a time when the onset of a business recession was typically followed in a few months by a decline in the price level and in wage rates, or at least by a moderation of the rise. That is no longer the case. The business cycle is still alive, indeed too much so; but its inner response mechanism, which has never stood still, is now very different from what it was even ten or twenty years ago.

Failure to perceive this may be responsible for some shortcomings in our national economic policy. I doubt if we will bring inflation under control, or even get a satisfactory expansion going, without a major shift in economic policy.

2. The first requirement of a new economic policy is recognition of the uneasiness that characterizes the economic world—an uneasiness that is slowing the business recovery and which may continue to impede it in this year and next.

Consumers are saving at a high rate, partly because of existing unemployment and partly because of concern about what inflation is doing to their incomes and savings.

There is also a mood of hesitation among businessmen. They see their wage costs accelerating, they fear that they will be unable to raise their prices sufficiently to cover their higher costs, and—with profits already sharply eroded and interest rates high—many hesitate about undertaking new investments. It is important to note that business capital investment, in real terms, has been declining in the present economic recovery; this is very unusual for this stage of the business cycle.

3. As you already know, I have reluctantly come to the conclusion that our monetary and fiscal policies have not been working as expected; they have not broken the back of inflation, nor are they stimulating economic activity as expected. . . .

I am convinced that the restoration of confidence requires, above everything else, a firm governmental policy with regard to inflation. With this achieved, our economy can move forward towards full employment. Without it, we may stumble—perhaps stumble all the more if we try a more expansive monetary or fiscal policy.

4. What I recommend, as a way out of the present economic malaise, is a strong wage and price policy.

I have already outlined to you a possible path for such a policy—emphatic and pointed jawboning, followed by a wage and price review board (preferably through the instrumentality of the Cabinet Committee on Economic Policy); and in the event of insufficient success (which is now more probable than it would have been a year or two ago), followed—perhaps no later than next January—by a six-month wage and price freeze.

Some such plan as this is, I believe, essential to come to grips with the twin problems of inflation and unemployment.

A firm wage and price policy will not get at the basic problem—the abuse of economic power. But the kind of legislation that would be needed will not be passed in the next year or two.

The wage-price freeze was very popular at first, a frightening development, as the natural flow of economic variables in the economy was being stifled. The stock market logged one of its best days ever. A number of economists, published in the *Wall Street Journal* and in *Newsweek* argued meanwhile against the freeze. Friedman was quoted in *Newsweek* as saying: "[President Nixon] has a tiger by the tail. Reluctant as he was to grasp it, he will find it hard to let go."[24]

Political responses varied. A year earlier in August 1970, amid public concerns on rising inflation, the Democratic-controlled Congress had passed the Economic Stabilization Act, giving the president the authority to stabilize prices, rents, and wages, which was interpreted inside the administration as a sort of dare. To take a sample from the AP newswires on August 17, 1971, in the Washington, DC, area, it was unsurprising, for example, to see Virginia Democratic senator William Spong Jr. arguing,

(*From left*) Federal Reserve chairman Arthur Burns, Treasury secretary John Connally, President Richard Nixon, Office of Management and Budget director George Shultz, and Council of Economic Advisers chair Paul McCracken at Camp David in August 1971, the weekend before unveiling the president's New Economic Program wage and price controls. Participants were instructed not to tell their families where they were headed in order to maintain the secrecy (and the theatrical nature) of the discussions. Having lost the argument against controls, McCracken would resign following the president's announcement. *Ollie Atkins for the White House. George Pratt Shultz papers, Box 732, Hoover Institution Archives.*

"The measures taken by the President were long overdue.... I advocated a freeze on wages and prices 18 months ago and I commend the President for taking this action."[25] But the state's Republican governor, Linwood Holton, also agreed:

> [The President] has taken a bold and positive action. I am very optimistic about the impact this action will have on the economy. He is doing everything which both labor and business have recommended.... I look forward to a material improvement in the unemployment situation and a restoration of consumer confidence.[26]

Independent senator Harry Byrd was more measured, saying, "The sweeping economic actions and proposals advanced by President Nixon will require close study before firm judgment can be made. Drastic action was made inevitable by swelling government deficits encouraged by both the Congress and the President."[27]

Corporate sentiment was similarly mixed. Richard Reynolds Jr., president of the Reynolds Metals Company, enthusiastically described the 1971 freeze:

> A very good and forceful move at a critical time. . . . The President's program is going to help the economy generally and basic industries, including aluminum, in particular. His action was certainly warranted by the condition of the economy. We don't have the demand. . . . It is quite clear that the President deserves praise for the scope of this action.[28]

Claiborne Robins, chairman and chief executive officer of A. H. Robins pharmaceutical company, meanwhile looked ahead, with some doubt, stating, "I think this probably is a good move for the short term, but over the long run I doubt controls are either effective or desirable."[29]

At first, the wage-price freeze seemed to work, and it came at a time when inflation was in the process of declining and commodity prices were soft in the world markets. The freeze was inevitably followed up by explicit, compulsory wage and price controls, which turned out to be very intrusive in the economy. People were unable to change wages and prices without the consent of the so-called Price Commission or Pay Board. The controls were administered with enthusiasm by a Cost of Living Council headed by John Connally, secretary of the Treasury.

In the short term, the consumer price index, which measures inflation, declined and real GDP rose. All this led to the landslide reelection of President Nixon. But trouble lay ahead. The economy sputtered, and prices were a problem. The Cost of Living Council, the bureaucracy responsible for administering the controls, was intrusive. No wage or price change could take place without approval by the Pay Board or the Price Commission, so the gears of the economy ceased working in the normal and natural way that produces an efficient system.

CHAPTER THREE

Damage Control as
the Economy Fights Back

S HULTZ BECAME secretary of the Treasury on June 12, 1972. The exchange rate system was in disarray after Nixon closed the gold window, a promise to buy gold with dollars at a fixed price, on August 15, 1971. An effort to reconstruct a fixed exchange rate par value system, known as the Smithsonian Agreement, had failed. With Friedman's help, Shultz designed a floating rate system in the clothing of a par value system, which the Europeans wanted. That proposal was put forward as the US position at a World Bank–IMF meeting. While it was never exactly implemented, the result might be called a managed float system. The nut of it was that exchange rates were, in effect, decontrolled. In the end, that worked reasonably well. So that left us with a market-based international system for currencies and a wage and price control system in the United States.

Meanwhile, the Cost of Living Council, now headed by Don Rumsfeld and Dick Cheney, reported to Shultz. But pushback was growing. Shultz's friend J. Willard Marriott, chairman of the board of Marriott Hotels, wrote to Shultz on January 15, 1973:

To Dr. Milton Friedman
With appreciation and best wishes,

June 1971

Richard Nixon

George Shultz and Milton Friedman with President Nixon in the summer of 1971. *Milton Friedman papers, Box 115, Folder 4, Hoover Institution Archives.*

The red tape which has involved numerous accountants, lawyers, discussions and meetings to get a legitimate price increase is impossible to explain to you. There is only one really fair way our industry can be handled and that is to release us from wage and price control and let us operate within the range of our fair profit situation. Our business is so competitive not only with other competitors, but especially with the housewife. And without having to fight continuously with the government, we have our hands full even keeping alive in our competitive market.

It is, I think, extremely obvious without looking at the figures that it is extremely unfair to control the retail prices of industry without controlling the wholesale prices, especially in the food industry where tremendous increases have occurred almost weekly in farm and wholesale prices.[30]

Later that year, on November 2, 1973, then secretary of commerce Frederick Dent wrote to the president reporting on a meeting he had had with labor and business leaders:

The first comments were made by a business representative and were directed to the necessity for decontrolling wages and prices. The ensuing general discussion on this topic brought out the following points:

- The wage and price program is creating an undue number of shortages in the economy, which can only be restored through free market actions.
- Selective decontrol will not work because industries are too highly interrelated. A general decontrol action is favored.
- To successfully accomplish decontrol, the people must be informed of the effects of controls, the reasons for decontrol, and the anticipated market response, while at the same time asking for their support and forbearance in achieving the essential goal.[31]

Shultz, Rumsfeld, and Cheney all thought it was time to ease up on the controls. They knew that inflation was suppressed, not eliminated, by controls, so they warned the president that probably there would be a slight uptick as the controls were eased up. This became known as phase III of the wage and price control system (see appendix D), which abolished the Price Commission and the Pay Board in favor of "self-administration" by obligated firms. But when the uptick came, Nixon decided to reimpose controls.

One humorous sideline to all this was a discussion in which Herb Stein, chairman of the Council of Economic Advisers, said to the president, referring to the popularity of the earlier wage-price freeze: "Mr. President, you can't walk on water twice," to which President Nixon replied, "You can if it's frozen." It was clear then where policy was headed.

Facing public outcry and political pressure, President Nixon reimposed price controls in June 1973 through a new sixty-day price freeze, beginning phase IV of the program, while simultaneously warning the American public against becoming addicted to the tool. Shultz had to say to the president, "This is your call, but it's directly opposed to my advice and I think you are making a mistake. Under the circumstances, you need to find a new secretary of the Treasury."

President Nixon accepted the reality but asked Shultz to stay on since Soviet leader Leonid Brezhnev was coming to Washington. Shultz had managed the US-Soviet economic relationship. That turned out to be very educational and useful, particularly when Shultz returned to government service in 1982 as secretary of state. For example, while Nixon didn't care for economics, he wanted to bring Brezhnev to California for security talks, so he made it possible to take the Soviet economists to Camp David for a weekend. While there, the Soviet economists engaged in lots of informal talk among themselves about their country's economy. They knew Shultz didn't speak Russian, so they talked freely. They didn't seem to realize that the US interpreter was also in the room, and he later gave a candid view of how poorly the Soviet experts thought their economy was working. This ended up influencing Shultz's views of Soviet bargaining positions on matters ranging from human rights to arms control a decade later. By March 1974, almost nine months after asking to resign, Shultz left the Treasury Department.

Nixon paid a price for his decision. Many years later, in a column published in the *New York Times* on July 7, 2003, nine years after Nixon's death, Bill Safire, a former speechwriter for Nixon, wrote that he had contacted Nixon "in purgatory, where he is still being cleansed of his sin of imposing wage and price controls."[32]

The political and economic damage caused by the wage and price controls lasted long after Nixon resigned on August 9, 1974. For one thing, the Fed under Arthur Burns increased money growth. This caused inflation to rise and ultimately brought on a recession in 1974–75. President Ford later established a new Council on Wage and Price Stability, which continued to intervene in price and wage setting, and he started a strange policy to reduce inflation.

Despite objections from his chief of staff, Donald Rumsfeld, Ford decided to give a speech before a joint session of Congress and announce a new "Whip Inflation Now" campaign, with a WIN button that people could wear to show that they intended

John Taylor and President Ford, January 1977. *The White House, courtesy John Taylor.*

to fight inflation through voluntary pledges to carpool or start a home vegetable garden. Ford simply rejected Rumsfeld's advice, saying, "Don, I think it is a good program."

In the speech, Ford said the pin was a "symbol of this new mobilization, which I am wearing on my lapel. It bears the single word WIN. I think it tells it all. I will call on every American to join in this massive mobilization and stick with it until we do win as a nation and as a people."[33] As Friedman predicted, with the Fed gunning the money supply, inflation remained high, backyard gardens or not, and the new intervention went nowhere.

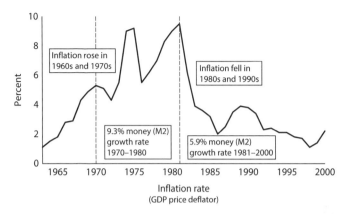

CHART 1. CHANGES IN MONETARY SUPPLY AND INFLATION. Although President Nixon's New Economic Program wage and price controls, enacted in August 1971, were effective in temporarily suppressing the inflation rate, economic fundamentals such as monetary supply growth, which is influenced by interest-rate decisions, were not reined in. Milton Friedman wrote that Nixon and Fed chairman Arthur Burns had "caught a tiger by its tail." As soon as those artificial controls were lifted, inflation quickly returned to its previous trajectory. And it resumed its upward march later in the 1970s as monetary supply growth remained loose, with "too many dollars chasing too few goods," as the saying goes. As the average rate of M2 growth, which measures the supply of money across the economy, gradually declined through the 1980s, inflation fell. *Data source: US Bureau of Economic Analysis; US Federal Reserve.*

Not all of the president's economic advisers were pushing these interventionist ideas. John Taylor arrived as a senior economist at the Council of Economic Advisers in the summer of 1976. Alan Greenspan, the chair of the council, was by no means an interventionist in his approach to economics. Indeed, the 1977 *Economic Report of the President* was quite noninterventionist. It included, for example, the mantra that "tax reform should be permanent rather than in the form of a temporary

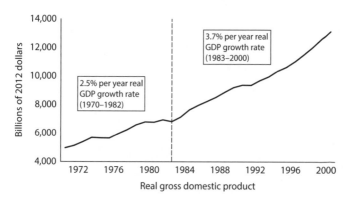

CHART 2. AN INFLECTION IN REAL GDP GROWTH. Compare the inflection in the inflation rate in the previous chart with this timeline of the GDP. The abysmal performance of the economy in the 1970s and the restoration of sanity at the start of the 1980s is instructive, as it can tell us something about what works and what does not work. It also shows the need to weather short-term economic fluctuations, which can spoil the politics, in the pursuit of a broader goal. As the next chapter demonstrates, this takes leadership. *Data source: US Bureau of Economic Analysis.*

rebate"[34] to emphasize that fiscal policy, as well as monetary policy, should be steady and predictable. But by that time the idea that government just had to "do something," which was ushered in with wage and price controls, had gained momentum and had spread to other areas of economic policy. The deviation from good, tried-and-true economic policy continued through the 1970s.

CHAPTER FOUR

Reversal and Reform

Good Economics Generates Good Policy

MILTON FRIEDMAN, a colleague and friend of ours at the University of Chicago and the Hoover Institution, was always a source of good advice. During a talk he gave to the O'Hare International Bank Executives' Club on March 14, 1974, Friedman described the widely different directions economic policy was then taking. Recall that the New Economic Program unveiled by Nixon after the August 1971 Camp David meeting had two major elements: One was domestic wage and price controls. The other was the end of foreign central banks' ability to convert their US dollar holdings to gold at a fixed rate (at the time $35 per ounce). Earlier that summer, West Germany had withdrawn from the Bretton Woods Agreement (the 1944 monetary order agreed upon by the major democracies; see appendix D), complaining of US dollar overvaluation. Thereafter, foreign governments began redeeming their US dollar holdings for gold as the dollar fell in value against other global currencies. Nixon knew he had to close the gold window, as Fort Knox wouldn't be able to continue supporting the demand for redemptions. And in doing so, as Friedman describes,

Milton Friedman and George Shultz with economist (and Nobel laureate) George Stigler at the East European Conference, May 8–10, 1991. Their panel was titled "Out of the Red: Economic Transition in Eastern Europe." *Hoover Institution on War, Revolution and Peace records, Box 3375B, Folder "Shultz, George," Hoover Institution Archives.*

John Taylor and Milton Friedman in San Francisco, August 18, 2006. Screen capture from a teleconferenced panel debate on monetary policy for the Festschrift conference in honor of Canadian economist David Laidler. Taylor recalls, "While walking up to the studio desk for the broadcast, Friedman tripped over a riser hidden in the shadows—glasses, papers all went flying. I was mortified. But Friedman was unflappable. He stood up, and with no ill will apparent at all, brushed himself off, and calmly sat down to begin the interview." *Courtesy John Taylor.*

in a somewhat messy way he actually began to move the world away from the fixed exchange rate system of Bretton Woods and toward one of floating, fiat money currencies relying on market dynamics. Friedman interrupted his prepared remarks to say the following:

On the radio this morning, I heard an announcement about the resignation of Secretary George Shultz, who is leaving the Treasury as of early May. Since George is one of us from Chicago, and more closely from the University of Chicago, I thought it might be appropriate to say a word about the extraordinary influence he has exerted. . . . I remember very well one speech of his (this was after price and wage controls had been in existence for some time), which started out, "There is a rumor going around that I am against price and wage controls. That rumor is true."

One interesting feature of George Shultz's performance is brought out by noting that within the past three years—beginning on August 15, 1971—we have had a dramatic experiment which no economic scientist could have improved upon, even if he had deliberately gone about to set one up. On August 15, 1971, President Nixon moved in opposite directions with respect to the international sphere and the domestic sphere.

With respect to the international sphere, Nixon closed the gold window and ended the commitment which the United States had to buy gold, or sell gold to foreign governments, at a fixed price. The ending of this commitment paved the way for the establishment of a system of floating exchange rates. It paved the way for a free market in the price of currency. It paved the way for ending controls on the price of the dollar in terms of the mark, franc, yen, and the pound. So, this measure opened the way to a free market, where there was none before.

At home, President Nixon moved in the opposite direction. He froze prices and wages, and that was followed—as you all know—by Phases 1, 2, 3, 4, 5, 6, 7, 8, and 9! Although some of those are in the future, prediction is easy in that area. Here was a beautifully controlled experiment. Compare the operation of free prices in the international market and controlled prices at home. It is hard to conceive of an experiment the results of which could have been clearer. In the international area, we have had a great success. In the domestic area, we have had a great failure. There is hardly a person today who will not recognize that price and wage controls meant more inflation—not less—that they meant disruption, distortion, a reduction in output, and an increase in the power of the government over the individual. Price and wage control has been a fiasco, and the administration is now proposing that it be ended when the present authority expires.

On the other hand, if you look abroad, the policy has been a great success. It's gotten much less news, much less attention, only because failure is always more newsworthy than success. Suppose we had continued with the system of fixed exchange rates. There is little doubt what would have happened. When the Arab-Israeli war broke out, and when the oil embargo came on, there would have been a major international financial crisis. We would have had tremendous flows of funds from the mark, pound, yen, and franc into the dollar. Central banks would have lost, or gained, billions of dollars of reserves overnight. There is little doubt that the foreign exchange markets of Europe would have been closed, little doubt that the headlines would have reported that the central bankers were retiring to some pleasant place like Bermuda to consider the world crisis, and to "solve the problem."

None of that happened. Why not? Because there was a free price system operating. These pressures were absorbed, not by massive flow of funds but by the price of the dollar in terms of other currencies rising by something like 5 percent to 10 percent, and then afterwards receding a little. The market worked just the way a starry-eyed theorist like myself, who has been proposing flexible exchange rates for many years, has said it would. The results were not those that were feared by the practical bankers like David Rockefeller, who were fearful of letting the market work abroad.[35]

Rockefeller had been chairman and CEO of Chase Manhattan and was a noted business community emissary on matters of international economic policy in the United States and abroad. Friedman continued:

I remember very well an episode some seven or eight years ago when I testified at the Joint Economic Committee on the question of flexible versus fixed exchange rates. It just happened that David Rockefeller was testifying the same day, along with his then economic adviser, Paul Volcker. David Rockefeller's position was that it was essential to maintain fixed exchange rates abroad; that there would be chaos in the world if the market controlled the price of foreign exchange. He would not say that today, because experience has demonstrated very clearly that precisely the opposite is true. The free market was able to meet a major crisis in the international sphere without serious difficulty, while the controlled market at home produced a major crisis without any difficulty.

Now I come back to George Shultz. The free market in exchange rates owes more to George Shultz's influence than to that
of any other individual. From the beginning, he consistently and
systematically pressed for flexible exchange rates, for allowing a
free market to operate. I may say that one of his most difficult op-
ponents was the banking community in the United States, including the Federal Reserve System. The Federal Reserve System was
consistently opposed to letting the exchange rates go free. They
wanted to intervene in the market, to control it. The Treasury, on
the other hand, consistently stood against it, and it is fortunate for
this country that on the whole George Shultz triumphed, though
unfortunately there has been some intervention in the market by
the Federal Reserve. The tragedy for this country is that his advice
and his position was not followed in the domestic area.

Friedman went on to describe how well-meaning attempts
by a variety of respected American business leaders to "do
something" on economic matters of direct interest to them
had unintended consequences given market fundamentals.
In the energy sector, for example, domestic oil price controls
(which by phase IV of the program had been capped at $4.25 per
barrel) were scheduled to gradually rise over six months toward
the world oil price. This plan was of course disrupted by the
1973 Arab oil boycott, which sent up global prices faster than
the US domestic oil prices were rising to meet them. Rockefeller
and others had been proponents of fixed exchange rates during
a period in which the Federal Reserve also limited the rates
that banks could offer on deposit accounts. These two financial
straitjackets led to the development of a largely London-based

"Euro-dollar" market in which European banks, which were not subject to such rate limitations, would actually make loans or accept deposits denominated in US dollars. Friedman saw the folly of the situation:

It's hard to realize how much better off we would be today if we had never embarked on the ill-fated attempt to control prices and wages. The blame is not easily placed for that attempt. I said a moment ago that I had an enormous amount of sympathy for the people at this table, but they have brought many of their troubles on themselves. The president of Union Oil Company, Fred Hartley, was one of the chief proponents of price and wage controls in 1971. How short-sighted men can be when they get outside the area of their own knowledge. The businessmen in 1971 who thought that price and wage control would consist of wage control without price control simply demonstrated their inability to count the number of voters. The so-called fuel and energy crisis today dates back to that period. Fred Hartley helped bring it on, not because he intended to, and not because he wanted to. I am not attributing blame in any other sense than the inability to see what the consequences would be.

Just in the same way, David Rockefeller and his fellow bankers in New York brought on the growth of the Euro-dollar market in Europe and were fundamentally responsible for driving inter-national financial business away from New York to London. The insistence on fixed exchange rates led to the exchange controls that President Johnson imposed, and which earlier Kennedy had imposed. Those exchange controls, together with Regulation "Q" limits on interest rates that could be paid to depositors, drove international financial business to Europe and caused the Euro-dollar market to flourish. This is another example of extraordinary short-sightedness. It was not deliberate—they didn't intend to drive business away, but they did. George Shultz also achieved the objective of ending all those exchange controls. . . . I don't know who will succeed him, but I was saying to someone that I hope he will be succeeded by Bill Simon, because if Bill Simon is taken away from the Federal Energy Office, where he is doing harm, and moved to the Treasury, he might do some good.

You know, an old teacher of mine, who was a mathematical statistician, once wrote a report on the teaching of statistics. In this report, he said, "Pedagogical ability is a vice rather than a virtue, if it is devoted to teaching error." That is a profound thought. Ability is a vice, not a virtue, if the able man is doing something bad. Mr. Simon is an able man, but the more poorly administered the Federal Energy Office is, the better the country will be.

By the time Friedman made these remarks in the spring of 1974, Nixon's formal wage and price controls had been scaled back under (the final) phase IV of the program: mandatory controls, which had covered 44 percent of the consumer price index (CPI) basket of all prices as of August 1973, had been scaled back to cover just 12 percent of CPI basket prices. Energy was one area in which price controls would persist for many years. Authority for oil and refined petroleum products' allocations and price controls was transferred to the newly created Federal Energy Office, headed by Shultz's former Treasury deputy William Simon, the implementation of which had led to rationing (hence Friedman's interest in Simon changing jobs, which he did two months later).

Friedman spoke further about the short-term versus long-term economic outlook:

Let me turn to what I had originally intended to talk about, which was, first of all, the short-term outlook and, later, the longer-term perspective. Insofar as the short-term outlook is concerned, we are clearly in an economic slowdown, recession, decline, or weakness—whatever you want to call it. I think the semantic issue is of no importance whatsoever. It couldn't matter an iota to anybody in this country whether real national output goes down by one-tenth of 1 percent a year or goes up by one-tenth of 1 percent a year. Yet, by some definitions of what is, or is not, a recession, that would be considered as making the difference between a slowdown and a recession. What is clear is that the growth of real output peaked sometime in the early part of 1973; that it has been going down erratically since; and that it is likely to continue to go down for some time. Unemployment reached a trough sometime in late 1973; has gone up moderately since then; and probably will go up some more. The ability of economists—or I may say of anyone else—to predict short-term changes is not great. Any statements made about the future course of events are bound to be subject to wide margins of error. About the only statement I think you can make with great confidence for the next six to nine months is that there is little prospect that the current decline will turn into a major recession, or depression. That is just not in the picture.

In a way, Friedman had actually predicted the recession—which was already under way when he made these remarks in Chicago—three years earlier in his October 1971 *Newsweek*

President-elect Ronald Reagan with his Coordinating Committee on Economic Policy advisers (*left to right*) Walter Wriston, Milton Friedman, Darrell Trent, George Shultz, Alan Greenspan, Paul McCracken, and others in Los Angeles, November 16, 1980. *Bettman via Getty Images.*

column, writing, "The most serious potential danger of the new economic policy is that, under cover of price controls, inflationary pressure will accumulate, the controls will collapse, inflation will burst out anew, perhaps sometime in 1973, and the reaction to the inflation will produce a severe recession."[36] The recession would continue into President Ford's tenure the following year, but economic performance and other fundamentals remained patchy through the decade. When Ronald Reagan was elected president in 1980, the real GDP was a little negative and the CPI was in the teens. There was residue from the controls mentality that had continued in a somewhat abated form through the Ford and Carter administrations. Amazingly, when President Reagan took office in January 1981, remnants of the bureaucracy set up to handle wage and price controls were still in existence. Everyone remembered Carter's gas lines as price controls in energy led to shortages during the 1979 Iranian Revolution.

Shultz chaired Reagan's Coordinating Committee on Economic Policy during the primaries and election and, for a while, after Reagan took office. This group presented a report to him in Los Angeles before his inauguration. The report went well beyond wage and price controls in recommending actions to improve monetary policy, to reform the tax code, and to counter the rising trend toward more government regulation. The report put into administrative form many of the ideas that Reagan had talked about during the campaign. Many of the recommendations in the report were followed, including the reduction in costly regulations, as illustrated in chart 3. The report is included as appendix C.

Among many other things, the committee advocated, "The current regulatory overburden must be removed from the economy. Equally important, the flood of new and extremely burdensome regulations that the agencies are now issuing or planning to issue must be drastically curtailed." In addition, the committee said, "An important step to demonstrate your determination to rely on markets would be the prompt end of wage and price guidelines and elimination of the Council on Wage and Price Stability."

At the outset of the Reagan administration, Paul Volcker was chairman of the Federal Reserve Board. In nominating Volcker before the election, Carter had chosen someone who could get the economics right, capping a decade spent trying everything but. It was Reagan, though, who saw how good economics at the Fed would support good policy overall. Earlier, he had been under secretary of the Treasury, so Shultz knew Volcker and

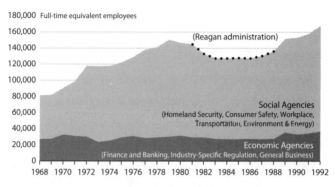

CHART 3. FEDERAL REGULATORY AGENCY STAFFING LEVELS. One way to approximate the level of regulation in the US economy is through the activity levels of relevant federal agencies. This chart shows full-time equivalent employee counts; agency funding levels would show a similar trend. On this count, the Reagan administration stands out for its deregulatory activities, including completing the phaseout of remaining wage and price controls, enacting two major tax cuts, and promoting a greater reliance on markets in place of government controls to meet social objectives. This led to higher economic growth. *Data source: Weidenbaum Center, Washington University and the Regulatory Studies Center, the George Washington University. Derived from the Budget of the United States.*

greatly respected him. He was doing what Arthur Burns had said would not work. The federal funds rate was high, as was the CPI. Gradually, both came down, as charts 1 and 5 show. During this period, people kept running into the Oval Office, saying, "Mr. President, he's going to cause a recession. We're going to lose seats in the midterm election." Volcker said that on many occasions the press served up questions to the president that invited him to criticize the Fed, but Reagan never took the bait. In other words, the president put a political umbrella over the Fed, taking a long-term point of view that we had to get inflation under control if we were to have a strong economy.

Years later, in the year 2000, Friedman looked back and assessed the inflationary experiences of the 1970s and the 1980s and the different approaches of Nixon and Reagan. It was part of an interview by John Taylor and started with Taylor asking Friedman about "what caused the Great Inflation of the 1970s and what caused its end." The answer recalled the low inflation before Kennedy's election and the suggestions in the early 1960s for wage and price guideposts that Friedman had opposed. The interview continued:

Friedman: Yes, the Great Inflation. The explanation for that is fundamentally political, not economic. It really had its origin in Kennedy's election in 1960. He was able to take advantage of the noninflationary economic conditions he inherited to "get the economy moving again." With zero inflationary expectations, monetary and fiscal expansions affected primarily output. The delayed effect on prices came only in the mid-sixties and built up gradually. Already by then, Darryl Francis of the St. Louis Fed was complaining about excessive monetary growth. Inflation was slowed by a mini-recession but then took off again when the Fed overreacted to the mini-recession. In the seventies, though I hate to say this, I believe that Arthur Burns deserves a lot of blame, and he deserves the blame because he knew better. He testified before Congress that, if the money supply grew by more than 6 or 7 percent per year, we'd have inflation, and during his regime it grew by more than that. He believed in the quantity theory of money, but he wasn't a strict monetarist at any time. [See appendix D.] He trusted his own political instincts to a great degree, and he trusted his own judgment. In 1960, when he was advising Nixon, he argued that we were heading for a recession and that it was going to hurt Nixon very badly in the election, which is what did happen. And Nixon as a result had a great deal of confidence in him.

From the moment Burns got into the Fed, I think politics played a great role in what happened. So far as Nixon was concerned, there is no doubt, as I know from personal experience. I had a session with Nixon sometime in 1970—I think it was 1970, might have been 1971—in which he wanted me to urge Arthur to increase the money supply more rapidly (laughter) and I said to the president, "Do you really want to do that? The only effect of that will be to leave you with a larger inflation if you do get reelected." And he said, "Well, we'll worry about that after we get reelected." Typical. So there's no doubt what Nixon's pleasure was.

Taylor: Do you think Burns was part of the culture of the times in that he put less emphasis on inflation, or that he was willing to risk some inflation to keep unemployment low, based on the Phillips curve?

Friedman: Not at all. You read all of Arthur's writings up to that point and one of his strongest points was the avoidance of inflation. He was not part of that Keynesian group at all. In fact, he wrote against the Keynesian view. [See appendix D.] However, it did affect the climate of opinion in Washington, it did affect what activities of the Fed were viewed favorably and unfavorably, and therefore it did affect it that way, but not through his own beliefs of the desirability of inflation.

Taylor: Another thing that people say now is that Burns was as confused as other people were about potential GDP, and that he thought the economy was either below capacity or that it was

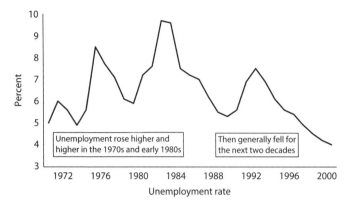

CHART 4. RISING AND FALLING UNEMPLOYMENT RATES. Contrary to some economic arguments made at the time, history shows that the unemployment rate actually rose through the period of high inflation in the 1970s. Both unemployment and inflation then fell beginning in the early 1980s. *Data source: US Bureau of Labor Statistics.*

capable of growing more rapidly than it was. Do you think that was much of a factor?

Friedman: I don't think that was a major factor. I think it may have been a factor.

Taylor: Mainly political?

Friedman: Yeah.

Taylor: What about the end of the Great Inflation? It lasted beyond Burns's time. We had G. William Miller and then Paul Volcker.

Friedman: Well, there's no doubt what ended it. What ended it was Ronald Reagan. If you recall the details, the election was in 1980. In October of 1979, Paul Volcker came back from a meeting in Belgrade, in which the United States had been criticized, and he announced that the Fed would shift from using interest rates as its operating instrument to using bank reserves or money. Nonetheless, the period following that was one of very extreme fluctuations in the quantity of money. The purpose of the announcement about paying attention to the monetary aggregates was to give Volcker a shield behind which he could let interest rates go. . . .

They did step on the brake, and in addition, sometime in February 1980, Carter imposed controls on consumer credit. When the economy went into a stall as we were approaching the election, the Fed stepped on the gas. In the five months before the election, the money supply went up very rapidly. Paul Volcker was political, too. The month after the election, the money supply slowed down. If Carter had been elected, I don't know what would have happened. However, Reagan was elected, and Reagan was determined to stop the inflation and willing to take risks. In 1981, we got into a severe recession. Reagan's public-opinion ratings went down, way down. I believe no other president in the postwar period would have accepted that without bringing pressure on the Fed to reverse course. That's the one key step: Reagan did not. The recession went on in 1981 and 1982. In 1982, finally Volcker turned around and started to raise the money supply, and at that point the recession came to an end and the economy started expanding.

Taylor: Your explanations of both the start and end of the Great Inflation are very much related to changes in people in leadership positions, as distinct from changes in ideas. What you seem to be saying is that it was mostly Burns, Nixon, Reagan. Could you comment on that a little bit?

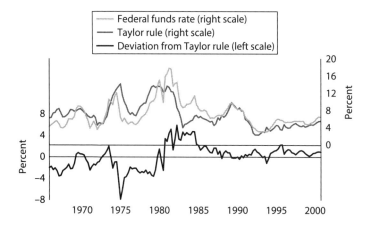

CHART 5. FEDERAL FUNDS RATE VS. TAYLOR RULE. We can use the Taylor rule to illustrate comments made by Milton Friedman about monetary policy choices made in the 1970s versus the 1980s. (For more information about the Taylor rule, see appendix D.)

The graph shows what the interest rate would have been under the Taylor rule along with the actual interest rate, the federal funds rate. The data are "real time," or what was available to the Fed at the time.

The deviation between the actual rate and the rule is plotted in the lower part of the graph. The deviation is large and negative in the 1970s, especially in the late 1970s. Inflation was high and variable, and output fluctuations were large during this period.

Policy then changed. A positive deviation in the early 1980s was as large as the negative deviation in the 1970s, but this was the transition to a new policy. During the transition—a period of disinflation—the interest rate went above the rule as the Fed brought inflation down. Following this transition, there were nearly two decades during which there was virtually no deviation. Economic performance was very good during this period.

Source: John H. Cochrane, John B. Taylor, and Volker Wieland, "Evaluating Rules in the Fed's Report and Measuring Discretion" (discussion paper prepared for presentation, Strategies for Monetary Policy conference, Hoover Institution, Stanford, CA, May 3, 2019). Based on Alex Nikolsko-Rzhevskyy, David H. Papell, and Ruxandra Prodan, "Policy Rules and Economic Performance" (white paper, December 11, 2018).

President Reagan meeting with President's Economic Policy Advisory Board members (*left to right*) Arthur Laffer, former Office of Management and Budget director James T. Lynn, George Shultz, incoming Council of Economic Advisers chair Murray Weidenbaum, and Congressman Jack Kemp in the Cabinet Room, February 10, 1981. *Courtesy Ronald Reagan Library.*

Friedman: I may be overemphasizing Burns's role. I certainly am not overemphasizing Reagan's. And again, in both cases I feel I have personal evidence. I was one of the people who talked to Reagan, and there's no question that Reagan understood the relation between the quantity of money and inflation. It was very clear, and he was willing to take the heat. He understood on his own accord, but he also had been told so, that you could not slow down the inflation without having a recession.

Taylor: In the first case, a president didn't take your advice, and in the second case, a president did take your advice.

Friedman: Correlation without causation. They were different characters and persons. Nixon had a higher IQ than Reagan, but he was far less principled; he was political to an extreme degree. Reagan had a respectable IQ, though he wasn't in Nixon's class. But he had solid principles, and he was willing to stick up for them and to pay a price for them. Both of them would have acted as they did if they had never seen me or heard from me.[37]

CHAPTER FIVE

Orthodox Policies and Strategic Thinking Work

S O WHAT lessons can we draw from all this experimentation in public policy? Above all, the main lesson is that orthodox policies work, and deviating from them can lead to trouble.

President Nixon, it might be said, did us a great favor by demonstrating that wage and price controls do not work. He imposed them on the economy and arranged to have them administered by talented people such as John Connally, John Dunlop, Don Rumsfeld, and Dick Cheney, among others. So he gave the country a lesson that, even with high talent at the helm, this approach doesn't work.

We also must be careful about the argument, as in the Arthur Burns letter, that the economy is a mess and doesn't work right anymore, so classical methods won't work. When you reach that stage of the argument, you almost inevitably reach for something different—in this case, wage and price controls.

As is likely to be the case, the analysis proved to be wrong. The prescription of drastic change proved to be wrong, and Paul Volcker in his role as chairman of the Federal Reserve used classical policies that Burns said would not work.

They did work. It took a while for inflation to come under control, but once it did, the economy took off. By the end of 1982, inflation was substantially reduced and stabilized, and everyone could see it was going to stay that way. The last remnants of the controls were killed off. In 1983, the economy took off like a bird. Economic expansion with stable, low inflation took place through orthodox policies. You deviate from these policies at your peril.

This lesson goes well beyond avoiding wage and price guideposts and controls. The same lessons apply to the other changes that began in the 1980s, including tax and regulatory policies. The marginal rate of taxation, having been reduced by Presidents Kennedy and Johnson from 90 percent to 70 percent, was brought down to 50 percent by Reagan and then, in the 1986 Tax Reform Act, to 28 percent. This change passed in the Senate by a vote of 97 to 3, suggesting that politicians were gradually understanding that lower tax rates were stimulating to the economy. The regulatory burden was also lightened in the Reagan era. Everything changed.

All these changes were recommended in the report to President-elect Reagan from the Coordinating Committee on Economic Policy (see appendix C), which we quote from here:

Sharp change in present economic policy is an absolute necessity. The problems . . . are severe. But they are not intractable. Having been produced by government policy, they can be redressed by a change in policy.

The essence of good policy is good strategy.

The need for a long-term point of view is essential to allow for the time, the coherence, and the predictability so necessary for success. This long-term view is as important for day-to-day problem solving as for the making of large policy decisions. Most decisions in government are made in the process of responding to problems of the moment. The danger is that this daily fire-fighting can lead the policy maker farther and farther from his goals. A clear sense of guiding strategy makes it possible to move in the desired direction in the unending process of contending with issues of the day. Many failures of government can be traced to an attempt to solve problems piecemeal. The resulting patchwork of ad hoc solutions often makes such fundamental goals as military strength, price stability, and economic growth more difficult to achieve.

Consistency in policy is critical to effectiveness. Individuals and business enterprises plan on a long-range basis. They need to have an environment in which they can conduct their affairs with confidence.

The 1990s

This story of good policy turning bad and then back again to good—as in this report—is a recurring one. When you look at what happened after the 1970s and 1980s as covered in this book, going into the 1990s and all the way to the present, you see similar stories. John Taylor remembers calling Milton Friedman from Washington in 1990 during a stint at the Council of Economic Advisers. Friedman was at Stanford's Hoover Institution and Taylor had the assigned task of calling him and others to see how much support there was for President George H. W. Bush's "revenue enhancements," or tax rate increases, which would begin to reverse the tax reforms of the 1980s. Taylor didn't even

President George H. W. Bush with (*left to right*) Chair Michael Boskin and Richard Schmalensee and John Taylor from the Council of Economic Advisers, February 1991. *The White House, courtesy John Taylor.*

have to say why he was calling, let alone ask the question about support for the tax change, before Friedman realized that the call was not simply to say hello and had a purpose. Friedman then made his views very clear. His words of wisdom were short and sweet. He simply said, "The answer is no!" adding, "You better come back to Stanford right away, John. Washington is corrupting you."

While taxes were raised rather than held down in the Bush administration, most other pressures to move in an interventionist direction in the years immediately after Reagan left town were resisted. Economic policy in the 1990s continued to be more market based, especially in comparison to the interventionist wage and price controls of the 1970s. Regulatory reforms led to more competition and innovation. In his 1996 State of the Union address, President Bill Clinton famously said that "the era of big government is over," and the federal budget moved into balance by the end of the decade. Fiscal policy relied more on automatic stabilizers rather than discretion. President George H. W. Bush did propose a small economic stimulus, but this failed to pass Congress, as did President Clinton's proposed stimulus package.

Meanwhile, Clinton, his federalism-minded budget director Alice Rivlin, and the Gingrich-Dole Congress found common ground in the "devolution" of moving authorities for various federal programs such as welfare to the states, where they could be more flexibly administered based upon local conditions. Liberalization also continued on the international side, with trade agreements such as NAFTA (the North American Free Trade Agreement, which led to an integrated North American supply chain) and the World Trade Organization (which opened new export markets while exposing US firms to global competition).

Monetary policy focused on inflation, a big change from the 1970s. The 1990 *Economic Report of the President*[38] complimented the Fed under Alan Greenspan, who succeeded Paul Volcker as chair, for attempting "to develop a more systematic longer-run approach." Greenspan maintained this commitment to price stability through the 1990s. The Fed focused more on transparency in its decisions. For example, in the 1970s, decisions about interest rates were hidden within the Fed's statements about the money supply. But in the 1990s, the Fed announced its interest-rate decisions immediately after making them.

The 2000s

As the twenty-first century began, one may have hoped that these same policies would continue and be applied to other government programs such as Social Security and even health care. But that's not what happened. Pressures for intervention continued to mount. People seemed to forget that good economic policy was responsible for economic success.

Public officials from both parties apparently found the orthodox approach of the 1980s and 1990s to be a disadvantage. Some wanted to do more, perhaps to better deal with the business cycle or stimulate home ownership. The good economic performance of the 1980s and 1990s was taken for granted, people became complacent about the successes, and they failed to resist political pressures that can lead to bad economic policy. Policy shifted in a more interventionist direction.

Policy change did not occur overnight. Federal interventions to promote the housing market, for example, grew gradually at first. A temporary tax rebate was passed in 2001 while Taylor was back in government as under secretary of the Treasury, despite his objections and those of Glenn Hubbard, chair of the Council of Economic Advisers. It did not stimulate the economy, and perhaps it was an early warning sign of a shift in policy. Though one could claim that the rebate was a first installment on the more permanent 2001 tax cuts that passed Congress and were signed by President George W. Bush, the decision led Friedman to pronounce with regret that "Keynesianism has risen from the dead." Given what has happened since then, he was certainly right.

President George W. Bush with John Taylor in the White House, shortly after Taylor assumed duties as under secretary of the Treasury for international affairs, June 7, 2001. *The White House, courtesy John Taylor.*

Other signs of activism were seen in monetary policy. Between 2003 and 2005, the Federal Reserve held interest rates far below the levels that would have been suggested by monetary policy rules that had guided the Fed's actions in the previous two decades. As measured against the Taylor rule and other metrics of monetary policy, the deviation was nearly as large as it had been during the poorly performing decade of the 1970s. The Fed added statements that interest rates would be low for a "prolonged period" and would rise at a "measured pace," an intentional departure from the policies of the 1980s and 1990s.

That departure was intended to help ward off a perceived risk of deflation, but the extremely low interest rates during these years contributed to the development of the housing bubble. The deviation, it seems to us, helped bring on the great recession and global financial crisis of 2007–09. Housing played the central role in the financial crisis that flared up in 2007 and turned into a panic in the fall of 2008. Another temporary stimulus package of $152 billion was passed, and interventionism reached a peak with the massive government bailouts of Wall Street firms in 2008.

Following the panic, the government could have returned to the less interventionist policies that had worked in the 1980s and 1990s. Testifying before the US Senate in November 2008, Taylor strongly recommended such policies rather than temporary tax reductions, including a pledge to stop the tax rate increases that were then scheduled to occur in the near future.

Instead Washington doubled down on its interventionist policies. On the monetary side, the Fed engaged in an unconventional monetary policy, including massive quantitative easing, which involved large-scale purchases of mortgage-backed securities and Treasury bonds to increase the money supply. The administration under President Obama did not reverse this trend. Rather, "it accelerated it," as Taylor later wrote in *First Principles: Five Keys to Restoring America's Prosperity*. There were more interventions—from an even larger $862 billion fiscal stimulus in 2009, which included temporary rebates and credits as well as grants to state and local governments, and many targeted programs including "cash-for-clunkers" vehicle trade-ins and tax credits for first-time home buyers.

Today

There are always pressures to change policy. Frequently the pressures are rationalized by arguments that economic principles have changed in fundamental ways. Avoiding costly deviations from good economic policy making requires vigilance on the part of policy makers and an insistence on a robust, open analytical approach, such as in listening to civil society and those outside of government as the above example of Milton Friedman illustrates.

What do we learn from all this? Point number one is to recognize that there is a constant debate in the United States and elsewhere about various ways of intervening in the marketplace to achieve some desirable results. The political process can easily create a demand to "do something" about important problems. The ability to stay the course on a strategic policy comes under great pressure.

Tried-and-true policies work and get results if given a chance. This leads to point number two, which is that the importance of thinking strategically cannot be overstated. Short-sighted thinking leads to bad results. Long-run thinking, even though it may mean a short-term hit, is much more productive. Orthodox, commonsense policies that are known to have worked need support. For the Fed, that means, it seems to us, a public statement of strategy (call it a rule—even a Taylor rule) from which it deviates only after a full explanation of why. The burden of proof should be on the deviation.

We make these points because today the air is filled with ideas for massive changes in the economy. Maybe some of these changes are needed, but be careful. Remember the mess of the economy caused by intervention in the 1970s and the way that President Reagan's policies put the economy back on a steady course of growth and expansion.

So remember, orthodox policies work, and excessive interventions by government in the market-based operations of the economy cause problems, sometimes severe. Watch out for charges that the economy has changed, so economic policies must change too. Stay with a long-term strategy and keep tax rates and the regulatory burden under firm control. These are the principles that the lessons of the 1970s and the 1980s have proven.

Timeline of Key Events

January 1962—President Kennedy's Council of Economic Advisers recommends voluntary "guideposts" for wage and price setting in its annual report.

April 1966—George Shultz and Robert Aliber organize a conference of leading economists at the University of Chicago on price-wage guideposts. The conference proceedings are published in *Guidelines, Informal Control, and the Market Place.*

June 9, 1970—Secretary of Labor George Shultz states his opposition to wage and price controls at a press conference at the National Press Club.

August 15, 1970—Congress passes the Economic Stabilization Act of 1970, giving the president authority to stabilize prices, rents, and wages.

February 12, 1970—Fed chair Arthur Burns testifies before the Joint Economic Committee: "It is the considered judgment of the Federal Reserve Board that, under present conditions, monetary and fiscal policies need to be supplemented with an incomes policy."

March 10, 1970—Burns testifies before the Senate Banking, Housing, and Urban Affairs Committee: "From a practical viewpoint, we face a problem unknown to earlier generations—namely, a high rate of inflation at a time of substantial unemployment."

April 22, 1970—George Shultz, now director of the Office of Management and Budget, delivers "Steady As You Go" speech to the Economic Club of Chicago, arguing that "the basic strategy of economic policy and its current tactical implementation are generally on course."

June 22, 1971—Burns sends "personal and confidential" letter to President Nixon urging wage and price controls.

August 13, 1971—President Nixon meets at Camp David with senior economic advisers, including Arthur Burns, Treasury secretary John Connally, under secretary for international monetary affairs Paul Volcker, George Shultz, secretary of commerce Peter Peterson, Council of Economic Advisers chair Paul McCracken and future chair Herbert Stein, Caspar Weinberger, and William Safire.

August 15, 1971, and following—President Nixon announces his New Economic Policy with (1) an executive order imposing a ninety-day freeze on all wages and prices (excluding agricultural goods), (2) a 10 percent imported goods and services surcharge, and (3) the end of dollar-gold convertibility.

"Cost of Living Council," composed of cabinet members and other senior officials, chaired by Treasury secretary Connally, is established and meets for the first time that afternoon.

A group of a dozen economists publishes a *Wall Street Journal* editorial arguing against the freeze. Political, corporate, and media reception to the freeze is generally supportive.

November 14, 1971—Cost of Living Council starts phase II of the controls, which would stay in place through the 1972 election, with Donald Rumsfeld as the director. Seven-member Price Commission and fifteen-member Pay Board are established.

January 12, 1973, and following—Phase III of the wage and price controls is instituted. The Price Commission and the Pay Board are abolished in favor of "self-administration" by "obligated parties."

Economist John Dunlop is named the new director of the Cost of Living Council.

June 13, 1973—Nixon reimposes a sixty-day price freeze, beginning phase IV of the program. Wages are not frozen.

July 18, 1973—Phase IV freeze is ended early by Nixon after just thirty-five days. Price controls will be gradually scaled back from covering 44 percent of CPI basket prices in August 1973 to 12 percent of CPI basket prices by their end in April 1974.

February 6, 1974, and following—John Dunlop tells the Senate Banking Committee of the administration's plans to end price controls in April 1974 except in health care and petroleum. Authority for price controls on oil and refined products is transferred to new Federal Energy Office.

Congress allows the Economic Stabilization Act of 1970 to expire, ending authority for most wage and price controls.

August 24, 1974—Congress passes and president signs the Council on Wage and Price Stability Act, which tasks a committee of economists to review, on behalf of the White House, significant federal agency rulemakings that could prove inflationary on the US economy.

October 8, 1974—President Ford addresses a joint session of Congress to announce the Whip Inflation Now campaign of voluntary pledges to limit inflation, including carpooling or starting backyard gardens. The effort is abandoned by the end of the year.

March 29, 1975—President Ford signs bill with Keynesian temporary tax rebate.

December 22, 1975—The Energy Policy and Conservation Act directs the Federal Energy Administration to allow the price of crude oil to rise gradually and to remove refined petroleum product price restrictions, subject to congressional review.

January 18, 1977—President Ford's Council of Economic Advisers issues annual report concluding that 1975 temporary tax rebate did no good, with John Taylor writing section titled, "Tax reduction should be permanent rather than in the form of a temporary rebate."

April 11, 1978—President Jimmy Carter announces voluntary inflation "deceleration" guidelines, asking unions and business to hold increases below the level of the previous two years.

October 24, 1978, and following—The new guidelines are expanded to a wage increase ceiling of 7 percent.

President's Council on Wage and Price Stability is tasked to review the profits of the four hundred largest US corporations, with data requests enforceable by subpoena.

November 16, 1980—Coordinating Committee on Economic Policy, chaired by George Shultz, delivers economic strategy report to President-elect Ronald Reagan.

March 31, 1981—Remaining price controls on domestic oil and refined petroleum are ended.

August 13, 1981—President Reagan signs the Economic Recovery Tax Act of 1981, which enacts across-the-board income tax cuts and tax incentives for investment.

October 22, 1986—President Reagan signs the Tax Reform Act of 1986, which had passed the Senate 97–3, enacting further broad-based tax cuts.

February 6, 1990—President George H. W. Bush's Council of Economic Advisers applauds a "systematic" rather than "undisciplined, ad hoc approach to policy" followed "consistently over time."

March 12, 1990—*Wall Street Journal* publishes front page story about John Taylor: "Bush Economist Is Urging Hands Off."

November 5, 1990—President George H. W. Bush signs the Omnibus Budget Reconciliation Act featuring "enhanced revenues"—otherwise known as higher income tax rates.

January 23, 1996—President Bill Clinton in State of the Union address declares, "The era of big government is over."

July 2001—Most American taxpayers begin receiving $300–$600 "stimulus" rebate checks as part of the Economic Growth and Tax Relief Reconciliation Act of 2001. This precedes the May 28, 2003, Jobs and Growth Tax Relief Reconciliation Act's tax rate cuts.

April 2008—Taxpayers again begin receiving $300–$600 rebate checks from the IRS as part of the Economic Stimulus Act of 2008 following Fed chair testimony on the desire to stimulate the economy to avoid recession. Federal Housing Authority loan limits are also increased in a bid to stave off falling home prices.

February 17, 2009—Shortly after taking office, amid the depths of the housing crisis, President Barack Obama signs the $862 billion American Recovery and Reinvestment Act stimulus package.

George Shultz with President Reagan at a meeting of the President's Economic Policy Advisory Board, February 10, 1981. *Courtesy Ronald Reagan Library.*

Letter from Arthur Burns to President Nixon

June 22, 1971

Chairman of the Board of Governors
Federal Reserve System
Washington, DC 20551

June 22, 1971

Personal and confidential

Dear Mr. President:

Last week, at the Quadriad meeting [of cabinet economic officials], I presented a set of statistical tables that give a rather comprehensive view of price and wage developments since the beginning of 1969.

I drew one conclusion from this evidence, namely, that it is doubtful whether we have made any progress in moderating the pace of inflation. The figures on the behavior of consumer prices in May reinforce my fears.

I realize that more optimistic interpretations have been made—and are still being made, by others—both within and outside the Administration. I have tried hard to read the evidence their way; I find that I cannot do so.

In my judgment, some of us are continuing to interpret the economic world on the model of the 1940s and 1950s. In fact, the structure of the economy has changed profoundly since then.

There was a time when the onset of a business recession was typically followed in a few months by a decline in the price level and in wage rates, or at least by a moderation of the rise. That is no longer the case. The business cycle is still alive, indeed too much so; but its inner response mechanism, which has never stood still, is now very different from what it was even ten or twenty years ago.

Failure to perceive this may be responsible for some short-comings in our national economic policy. I doubt if we will bring inflation under control, or even get a satisfactory expansion going, without a major shift in economic policy.

Let me turn, first, to the question: Why are the old rules not working? Why is inflation continuing at a virtually undiminished rate in the face of substantial unemployment of men, machinery, plant and equipment?

The answer is to be found in the changed character of our labor and product markets. Here are a few of them:

1. Our success in moderating downturns in economic activity during the postwar period has reduced the concern of workers over the possibility of prolonged unemployment and the fears of businessmen of a protracted slump in sales. Wage and price decisions are now being made on the assumption that governmental policy will move promptly to check a sluggish economy.

2. The past decade has witnessed a vast growth of trade unionism in the public sector—mainly among State and

local employees but also Federal employees. We have had innumerable illegal strikes of public employees, among them the postal strike. These strikes have been preponderantly successful, in the sense that the employees got pretty much what they asked for. In this environment, both the trade union leaders and the rank and file have come to believe that the government lacks either the power or the will to resist their demands. They have therefore become progressively bolder in pushing wage and other demands.

3. During the past decade, we have also had a vast expansion of welfare programs. When subsidies are provided by government to strikers—and this is what we have been doing more and more—workers on strike can hold out longer. Business managers are fully aware of this, and they are therefore less inclined to take a strike than was the case ten or twenty years ago.

4. The Landrum-Griffin Act has weakened the power of trade union leaders relative to the rank and file membership. As a result, when the leaders have sought moderate wage settlements, they have frequently been thwarted by a rebellious membership.

5. Non-union enterprises have always been influenced in their wage policies by what the unionized establishments were doing. With trade unions becoming stronger and more influential, nonunion businesses now adjust more quickly to—or even anticipate—the wages being set in the unionized sector.

6. The persistence of inflation, which has been running a mad
   course since 1964, has by now made practically every par-
   ticipant in the economic process inflation-minded.

   a. Workers seek wage increases that will compensate for
      past effects of inflation on their real incomes and pro-
      vide some insurance against future price advances, be-
      sides providing for some improvement in their living
      standard.

   b. Businessmen mark up their prices even when demand
      is weak, partly because they expect their costs to go
      up, partly because they fear further erosion of their
      shrunken profit margins, and partly because they expect
      that their competitors—facing similar problems—will
      behave similarly.

   c. Buyers, whether in the consumer or the business sec-
      tor, have become accustomed to paying higher prices,
      and therefore offer less resistance to price advances
      than they would in an environment of stable price
      expectations.

   d. Businessmen, expecting to pay back in cheaper dollars,
      have become more inclined to borrow. Lenders, having
      similar expectations with regard to inflation, tend to
      hold out for higher interest rates. Every sign that infla-
      tion may be quickening leads to expectations of higher
      interest rates and thus reinforces these tendencies.

   e. In short, every part of the decision-making process in
      the business and financial world is now being shaped
      by inflationary expectations, even in the face of heavy
      unemployment.

We thus face an entirely new economic problem—one that our nation has never before had to face: namely, an inflation feeding on itself at a time of substantial unemployment.

Much of our economic thinking and economic policy has not yet caught up with the changes that have taken place in the structure of our economy. We continue to rely on monetary and fiscal policies that worked reasonably well ten or twenty years ago, unmindful of the profound changes in our economic environment.

The first requirement of a new economic policy is recognition of the uneasiness that characterizes the economic world— an uneasiness that is slowing the business recovery and which may continue to impede it this year and next. Consumers are saving at a high rate, partly because of existing unemployment and partly because of concern about what inflation is doing to their incomes and savings.

There is also a mood of hesitation among businessmen. They see their wage costs accelerating, they fear that they will be unable to raise their prices sufficiently to cover their higher costs, and—with profits already sharply eroded and interest rates high—many hesitate about undertaking new investments. It is important to note that business capital investment, in real terms, has been declining in the present economic recovery; this is very unusual for this stage of the business cycle.

As you already know, I have reluctantly come to the conclusion that our monetary and fiscal policies have not been working as expected; they have not broken the back of inflation, nor are they stimulating economic activity as expected.

What the economy basically lacks is sufficient confidence in the future. There is a widespread feeling that we are drifting, that inflation may be getting out of control. Worried, unhappy people will not act to enlarge their activity in the economic arena or any other. I am convinced that the restoration of confidence requires, above everything else, a firm governmental policy with regard to inflation. With this achieved, our economy can move forward towards full employment. Without it, we may stumble—perhaps stumble all the more if we try a more expansive monetary or fiscal policy.

The recent nervousness of financial markets supplies a warning that we must not overlook: every time we move, or appear to move, in a more stimulative direction, interest rates rise. Needless to say, this disconcerting behavior could frustrate our hopes with regard to residential building and State and local construction.

What I recommend, as a way out of the present economic malaise, is a strong wage and price policy.

I have already outlined to you a possible path for such a policy—emphatic and pointed jawboning, followed by a wage and price review board (preferably through the instrumentality of the Cabinet Committee on Economic Policy); and in the event of insufficient success (which is now more probable than it would have been a year or two ago), followed—perhaps no later than next January—by a six-month wage and price freeze.

Some such plan as this is, I believe, essential to come to grips with the twin problems of inflation and unemployment.

A firm wage and price policy will not get at the basic problem—the abuse of economic power. But the kind of legislation that would be needed will not be passed in the next year or two.

Also, a firm wage and price policy such as I'm suggesting may not restore full price stability. But it will at least clear the air, dampen inflationary psychology, bolster confidence, and perhaps leave elbow room for a more stimulative fiscal policy to hasten the decline of unemployment.

Let me say, finally, that my omission of any reference to a more stimulative monetary policy is deliberate. Monetary policy, I feel, has done its job fully.

The issues that I've touched on in this letter are clearly of the greatest importance. If you should want me to clarify any part of my thesis, do let me know.

Sincerely yours,
Arthur F. Burns

The Honorable Richard Nixon
The President of the United States
The White House
Washington, DC

Economic Strategy for the Reagan Administration

*A Report to President-Elect Ronald Reagan
from His Coordinating Committee
on Economic Policy*

November 16, 1980

Sharp change in present economic policy is an absolute necessity. The problems of inflation and slow growth, of falling standards of living and declining productivity, of high government spending but an inadequate flow of funds for defense, of an almost endless litany of economic ills, large and small, are severe. But they are not intractable. Having been produced by government policy, they can be redressed by a change in policy.

The Task Force reports that you commissioned during the campaign are now available. They contain an impressive array of concrete recommendations for action. More than that, the able people who served on the Task Forces are available to provide further detail and backup information to you or your designees. We all want to help, and you can count on enthusiastic and conscientious effort.

Your Coordinating Committee has reviewed the Task Force reports. With due allowance for some differences in view about particulars and relative importance, we have found that they offer a substantial base for action by you and the team you assemble. We focus here on guiding principles, on priorities and linkages among policy areas, and on the problem of getting action.

You have identified in the campaign the key issues and lines of policy necessary to restore hope and confidence in a better economic future:

- Reestablish stability in the purchasing power of the dollar.
- Achieve a widely shared prosperity through real growth in jobs, investment, and productivity.
- Devote the resources needed for a strong defense and accomplish the goal of releasing the creative forces of entrepreneurship, management, and labor by:
  - Restraining government spending.
  - Reducing the burden of taxation and regulation.
  - Conducting monetary policy in a steady manner, directed toward eliminating inflation.

This amounts to emphasis on fundamentals for the full four years as the key to a flourishing economy.

Guiding Principles
The essence of good policy is good strategy. Some strategic principles can guide your new administration as it charts its course.

- *Timing and preparation are critical aspects of strategy.* The fertile moment may come suddenly and evaporate as quickly. The administration that is well prepared is ready to act when the time is ripe. The transition period and the early months of the new administration are a particularly fertile period. The opportunity to set the tone for your administration must be seized by putting the fundamental policies into place immediately and decisively.

- *The need for a long-term point of view is essential to allow for the time, the coherence, and the predictability so necessary for success.* This long-term view is as important for day-to-day problem solving as for the making of large policy decisions. Most decisions in government are made in the process of responding to problems of the moment. The danger is that this daily fire-fighting can lead the policy maker farther and farther from his goals. A clear sense of guiding strategy makes it possible to move in the desired direction in the unending process of contending with issues of the day. Many failures of government can be traced to an attempt to solve problems piecemeal. The resulting patchwork of ad hoc solutions often makes such fundamental goals as military strength, price stability, and economic growth more difficult to achieve.

- *Central problems that your administration must face are linked by their substance and their root causes.* Measures adopted to deal with one problem will inevitably have effects on others. It is as important to recognize these interrelationships as it is to recognize the individual problems themselves.

- *Consistency in policy is critical to effectiveness.* Individuals and business enterprises plan on a long-range basis. They need to have an environment in which they can conduct their affairs with confidence.

- *Specific policies as well as long-term strategy should be announced publicly.* The administration should commit itself to their achievement and should seek congressional commitment to them as well. Then the public, as well as the government, knows what to expect.
- *The administration should be candid with the public.* It should not over-promise, especially with respect to the speed with which the policies adopted can achieve the desired results.

Seizing the Initiative

The fundamental areas of economic strategy concern the budget, taxation, regulation, and monetary policy. Prompt action in each of these areas is essential to establish both your resolve and your capacity to achieve your goals.

Budget

Your most immediate concern upon assuming the duties of the President will be to convince the financial markets and the public at large that your anti-inflation policy is more than rhetoric. The public, and especially the financial community, is skeptical and needs a startling demonstration of resolve. Many question whether you are serious about a sizeable cut in budget outlays. Credible FY 1981 and 1982 budgets which do that clearly and unambiguously would evoke an extraordinary response in the financial markets and set the stage for a successful assault on inflation and a decline in mortgage and other interest rates.

The FY 1981 budget will be almost four months along by the time you take office, and a FY 1982 budget will have been submitted for consideration by the Congress. There are now estimates of alarming increases in these swollen budgets. Prompt and strong action is necessary if these budgets are to be brought under control, as they must be. The nation can no longer afford governmental business as usual.

The formal budget alone is far from the whole story, though it is visible and important. Off-budget financing and government guarantees mount and expand programs through the use of the government's borrowing capacity, draining the nation's resources without being adequately recorded in the formal spending totals. In addition, the mandating of private expenditures for government purposes has gained momentum as the spotlight has illuminated direct spending. These mandates are also a clear call by government on the nation's resources. Efforts to control spending should be comprehensive; otherwise, good work in one area will be negated in another. And these efforts should be part of the administration's development of a long-term strategy for the detailed shape of the budget four or more years into the future.

The Weinberger Task Force has identified an extensive and promising array of areas for potential savings, but it will be up to your administration and the Congress to do the job. It takes top-notch people to do it. We recommend that:

- A Budget Director, permanent or *pro tem*, be chosen and set to work now.
- A small team from OMB be assembled explicitly to work with the newly designated Director.
- The Director's recommendations be a part of your discussion with Cabinet and sub-Cabinet appointees as these appointments are made.

Amendments calling for dramatic reductions in the FY 1981 budget should be submitted to the Congress within the first week of your administration. A thoroughly revised FY 1982 budget provides even greater opportunities for large further reductions, and this budget should be submitted as soon as possible.

Finally, it has become all too evident in recent years that current budget procedures are biased in an expansionary direction. The congressional budget process defined by the Budget Act of 1974 has failed to achieve its purpose of removing the "runaway" bias. We therefore recommend a presidential task force to develop new techniques which can help to rein in the growth of federal outlays. It should examine the presidential item veto, renewed presidential power to refrain from spending appropriated funds, and other initiatives to hold down spending. This task force should report to you within two months.

Tax Policy

Tax policy is properly the province of your Secretary of the Treasury. The making of that appointment should have a high priority so that important work can go forward. The Walker Task Force provides the materials needed to pose the issues to you in concrete form and to translate your decisions into a proposal to the Congress. This proposal should be presented early in the new administration in tandem with other key elements of your economic program. It should embody the main thrust of tax policy for the whole of your first term, not simply for the year 1981. We consider that the key ingredients should be your proposals for the Kemp-Roth cut in personal income tax rates, simplification and liberalization of business depreciation, and a cut in effective taxes on capital gains (see Innovation Task Force). Consistent with your proposals earlier this year, the effective date for these reductions should be January 1, 1981.

Other key proposals are tax incentives for the establishment of enterprise zones in the inner cities and such other items as tuition tax credits, reductions in the windfall profits tax, inheritance taxes and the taxation of Americans living abroad, and the restoration of restricted stock options.

Regulation

The current regulatory overburden must be removed from the economy. Equally important, the flood of new and extremely burdensome regulations that the agencies are now issuing or planning to issue must be drastically curtailed. The Weidenbaum Task Force sets out the needed blueprint for personnel selection, immediate administrative action, and legislation. Again, the key to action is a knowledgeable and forceful individual to develop and coordinate strategy and to form a team to carry it out. Such an appointment should be made promptly, with the expectation that the effort would carry forward through the transition for at least a year into your administration. Your appointee and his team should be located within the Executive Office of the President. Achieving regulatory reform will take informed, strong, and skillful work with the Congress as well as with those in charge of departmental and agency regulatory efforts. The person heading up this effort will require your continued, wholehearted support.

Many of our economic problems today stem from the large and increasing proportion of economic decisions being made through the political process rather than the market process. An important step to demonstrate your determination to rely on markets would be the prompt end of wage and price guidelines and elimination of the Council on Wage and Price Stability.

To advance the entire regulatory effort—both to galvanize public support and to strengthen the positions of administration appointees—we urge you to issue a message on regulatory reform in tandem with the budget and tax messages. The message should call upon state and local governments to launch similar regulatory reform efforts—as a few have already done.

Energy

The battle between government regulation and the private market is nowhere more apparent than in energy, where the market has a decisive comparative advantage. Governmental intrusion into energy production and use provides a glaring example of how regulation costs us all dearly. Alternatives to imported oil exist here in the United States. As the Halbouty Task Force emphasizes, market pricing and market incentives will accelerate the development of these alternatives just as surely as present regulations and the politicization of this field inhibit them. Its recommendations and the issues it poses for careful review cover the energy field in a comprehensive manner and deserve immediate attention.

We recommend, also, that you promptly exercise the discretion granted to the President to remove the price controls on crude oil and petroleum products rather than continue with the present calendar, which postpones complete decontrol until October 1981. This decisive action will eliminate at once the regulatory apparatus administering the entitlements program and discourage continued efforts by special interests to prevent or slow down decontrol and deregulation. Also, the Natural Gas Policy Act of 1978 should be repealed so that all natural gas prices are decontrolled. These measures are particularly urgent because the uncertainty of our critical Middle East oil supplies, dramatized by the Iran-Iraq war, makes it all the more necessary to get the earliest possible incentive effect of free market pricing.

The Synthetic Fuels Corporation is incompatible with the free market pricing of energy and should be promptly eliminated.

The Department of Energy has become a large and unmanageable institution with a variety of programs ranging from essential to useless. The essential functions should be transferred and the department eliminated. This should be the main task of your Secretary of Energy.

Monetary Policy

A steady and moderate rate of monetary growth is an essential requirement both to control inflation and to provide a healthy environment for economic growth. We have not had such a policy. The rate of monetary growth declined sharply in the early months of 1980 and rose rapidly in recent months. These wide fluctuations are adversely affecting economic conditions and may continue to do so into 1981.

The McCracken Task Force emphasizes that the attainment of a proper monetary policy deserves the very highest priority and that such a monetary policy can be achieved through effective use by the Fed of its existing powers. The Task Force also brings out the relationship of monetary policy to budgetary and other economic policies.

The Federal Reserve is an independent agency. However, independence should not mean lack of accountability for what it does. In practice, independence has not meant that the Federal Reserve is immune to presidential and congressional influence. The problem is how to assure accountability while preserving independence. We suggest that you:

- Request the Fed to state targets for monetary growth year by year for the next five years that, in its opinion, will end inflation. Influential members of relevant committees of Congress have already urged the Fed to specify such long-term targets.
- Assure the Fed that you will propose and fight for fiscal and other policies compatible with the elimination of inflation.
- Improve the procedures for coordinating Federal Reserve monetary policy with the economic policies of the administration and the Congress and support congressional efforts to monitor the Fed's performance and to recommend changes in the procedures that could improve performance.

With these fundamentals in place, the American people will respond. As the conviction grows that the policies will be sustained in a consistent manner over an extended period, the response will quicken. And a healthy US economy, as the Burns Task Force states, will restore the credibility of our dollar on world markets, contribute significantly to smoother operation of the international economy, and enhance America's strength in the world.

Organizing for Action

The activities of a wide variety of departments, agencies, and other units of government within the Executive Branch impinge on economic policy. But the flow of economic events does not recognize organization lines. The economy itself operates as a system in which constituent parts are linked, sometimes tightly. The combination of interwoven problems and disparate organizations means that, in the process of policy formulation and implementation, some people high in your administration must identify the central ideas and problems and devise a strategy and tactics for dealing with them. Your leadership is essential to this effort.

One arrangement that has worked well in the past is for the Secretary of the Treasury to be the chief coordinator and spokesman on economic policy, domestic and international. To carry out this mandate effectively, the Secretary should be one of your key staff members as well as a departmental head with a White House title and office. Since economic developments are often closely related to security, the Secretary should be a member of the National Security Council. For this coordinating role, an Economic Policy Board, with comprehensive membership, should be established; it should meet regularly and be the avenue through which economic issues come to your Executive of the Cabinet and to your desk. The Council of Economic Advisers might suitably provide the secretariat for this group.

Maintaining a Steady Course

Our final point is our most important one. The success of your economic policy will be a direct reflection of your ability to maintain a steady course over your full first term. Rough times will come and crises of one kind or another—some small, some of great moment—will arise. Sustained effort through these testing times means that public understanding and support are essential. Of equal and related importance is the understanding and support of the Congress.

This last task—gaining understanding and support of the Congress—is of crucial importance. As a result of the voting on November 4, the 97th Congress, we are convinced, will be more cooperative on economic and financial issues. That cooperation will be fostered if, during the transition, the Secretary of the Treasury (designate) consults intensively with key members of Congress on the design and implementation of your economic policies.

You have emphasized in your successful campaign precisely the strategy set forth in this document. In moving to implement it, you will be doing what the people voted for. Every effort must be made to maintain and broaden your base of support by improving public understanding and by close cooperation with the Congress. Cabinet officers and others in your administration can help in these tasks. Their ability to do so should be one important criterion in their selection.

At the end of the day, however, the burden of leadership falls on you: leadership to chart the course ahead; leadership to persuade that your course is the one to take; leadership to stay on course, whatever way political winds may blow. Through effective advocacy of the sharp changes so sorely needed, your leadership has brought us to this long-hoped-for opportunity at a critical moment for the nation. Your leadership can maintain this advocacy in the convincing manner necessary for a successful outcome.

Arthur F. Burns

Milton Friedman

Alan Greenspan

Michel T. Halbouty

The Honorable Jack Kemp

James T. Lynn

Paul McCracken

William E. Simon

Charls E. Walker

Murray L. Weidenbaum

Caspar W. Weinberger

Walter B. Wriston

George P. Shultz,
Chairman

The final signature page of members and the chairman of the Coordinating Committee on Economic Policy. *George Pratt Shultz papers, Box 1256, Hoover Institution Archives; individual task force reports also available at this location.*

Frequently Asked Questions

What are the differences among guideposts, guidelines, incomes policy, and controls on wages and prices?

These are all various measures that governments use to intervene in individual wage and price decisions in an economy. The measures were not invented recently. The Roman emperor Diocletian issued an edict in AD 301 that controlled wages and prices by setting a ceiling on what could be paid for all kinds of jobs and goods. And, of course, under central planning the government tries to set prices for most goods and services. The minimum wage is a requirement by the federal government and many state governments that firms cannot pay workers less than a certain amount.

The call by the Kennedy Council of Economic Advisers for guideposts in the 1962 *Economic Report of the President* set off a two-decade-long bout of government interference in price setting, which is the focus of this book. *Guideposts*, sometimes called *guidelines*, are often voluntary and apply to many goods and services, but they bring the government into interference with private-sector decisions with harmful effects, as described by Milton Friedman in this book. *Controls* usually refer to mandatory ceilings on prices and wages, such as freezes or requirements that the rate of change of prices or wages not go above a certain amount. *Incomes policy* is different from voluntary wage and price guidelines or guideposts in that it is mandated by law. One variant of incomes policy is "tax-based incomes policies" (TIPs), in which a tax is imposed on firms that raise prices more than stated on the controls. During the 1970s, price and wage increases had to be approved by the Price Commission or Pay Board.

To implement any intervention in wage and price setting, the government either compels or otherwise jawbones businesses into setting prices for goods and services below the levels that firms would otherwise aim for given expected market conditions. Similarly, the state directs organized labor leaders, who would otherwise seek to maximize employee wages through collective bargaining with corporate management, to moderate or postpone their wage demands and associated bargaining tactics.

What were the different phases of wage and price controls during the 1970s?

There were four phases with timelines as follows:

- Phase I (August 15, 1971–November 13, 1971): wages, prices, and rents are frozen.
- Phase II (November 14, 1971–January 11, 1973): wage and price increases must stay within guidelines monitored by the Cost of Living Council and administered by the Pay Board and the Price Commission; large firms must get approval and smaller firms must report any increases exceeding the guidelines.

- Phase III (January 12, 1973–August 12, 1973): wages and prices can be self-regulated; but an explosion of prices leads to a refreeze on June 13, 1973, for thirty-five days.
- Phase IV (August 13, 1973–April 13, 1974): Cost of Living Council tries to increase supply of goods across the economy even with price freeze. Agricultural goods, for example, are exempted in a bid to encourage production, but with retail food prices still under control, shortages result.

What is the Phillips curve?

The Phillips curve is a representation of the negative correlation between inflation and unemployment, which can be shown by a negatively sloped line on a graph with inflation on the vertical axis and unemployment on the horizontal axis; in other words, when inflation rises, unemployment drops. It is named after A. W. Phillips, the economist who first showed that such correlations existed in British data from 1861 to 1957. The Phillips curve was used in the 1960s and 1970s to justify a monetary policy that included higher inflation. People argued that higher inflation would lead to lower unemployment, that there was a long-run trade-off between inflation and unemployment. However, Milton Friedman showed that this was a fallacy: there was no long-run trade-off, and low unemployment and low inflation could exist simultaneously. Policy was much different in the 1980s largely because of Friedman's analysis.

What is meant by the term *market power*?

A firm has market power when it has few competitors that could undercut the firm's price and charge a lower price for the good or service. For example, a monopoly, by definition, is the only firm in the market and thus has no competitors; it is the only firm selling the product. Thus, a monopoly can charge a high price without any concern that another firm will undersell it. It has a great deal of power—that is, market power—to set the price in its market. Customers may not buy as much from the monopolist at a higher price, but no other firm will charge less. In markets where there are only a few firms—say two or three or four—the firms also have market power, but not as much as in a monopoly. In this case, firms may implicitly collude about the price, and thereby keep the prices higher than under competition.

Robert Solow refers to market power in this book as a source of upward pressure on inflation because firms with market power can raise their price without being checked by competitive pressures. Hence, he argues for guidelines to hold down the size of their price increases. Most economists agree that market power leads to the *level* of prices being higher than they would be in a purely competitive market. But many economists—including Friedman—add that this does not mean that *inflation* (the rate of change in prices) would be higher with more market power. Thus, there is room for disagreement, as the book shows.

What was the WIN campaign?

The WIN, or "Whip Inflation Now," campaign was one of the strange ways that President Ford continued to try to control wage and price inflation after President Nixon resigned. The campaign included a "WIN" button that people could wear on their lapels if they mailed in personal pledge cards to Washington, DC. Ford announced it during a speech to a joint session of Congress and he wore the WIN button during the speech. His idea was that firms and workers would not set higher prices nor ask for higher wages as part of a public service, even if it were not in their own interest. The new intervention effort went nowhere.

What are classical economic policy and Keynesian economic policy?

For many years, there has been a widely perceived contrast between Keynesian economic policy and classical economics. We use the word *Keynesian* to refer to policies that call for more interventionist approaches, such as wage and price guidelines and controls. *Classical* economics refers mainly to policies that call for more use of markets, including freely determined wages and prices rather than controls, and a cost-benefit approach to regulations. A very famous paper written in 1937 by Nobel Prize winner John Hicks, "Mr. Keynes and the Classics," not only gave a good exposition of what Keynes described in his 1936 book *The General Theory*, it also portrayed classical economists as following the view that only money mattered and that wage cuts were all that was needed to put an economy on track to full employment. In this book, we consider classicists as adherents to the sound monetary and fiscal policy of the kind introduced in the 1980s and 1990s. Effectively, we think of Bob Solow as being in the Keynesian school and Milton Friedman as being in the classical school, although, as in all terminology distinctions, this classification is not perfect.

What was the Bretton Woods system?

The summer of 1971 marked the end of a long-held system for governing the economic relations of the United States, Western Europe, Japan, Canada, and Australia, as first negotiated by those states at a conference in Bretton Woods, New Hampshire, in 1944, during the Second World War. Allied powers had sought to avoid the rapid currency swings of the 1930s by establishing a "par value," or fixed, exchange rate system to allow free convertibility of their respective currencies. These would all be tied to the US dollar as a reserve currency, which would in turn be fixed to the price of gold at $35 per ounce. The goal was for a predictable international currency regime that would encourage freer trade in the years following the war, and it led to the centrality of the US dollar in international commerce. The International Monetary Fund was established as a multilateral organization to help manage requests from member governments for deviations from these rates and to aid in debt crises.

This fixed exchange rate system began to break down during the late 1960s, as described in this book. On the one hand, the economic hegemony of the United States declined relative to other members as their economies grew, and some other global leaders grew resentful of the impact of US monetary choices on their domestic currencies. And by the summer of 1971, US deficit spending and domestic inflation led to complaints of dollar overvaluation against global currencies. West Germany subsequently withdrew from Bretton Woods, foreign governments began redeeming their own US dollar holdings for gold at the fixed rate, and the US Congress Joint Economic Committee released a report recommending devaluation of the US dollar. Nixon's closing of the gold window alongside the introduction of phase I wage and price controls spelled the beginning of the end of the Bretton Woods system and eventually led to today's largely free-floating market-based currency exchange rates, as described in Shultz's and Friedman's remarks in this book.

What is the quantity theory of money?

The quantity theory of money represents a relationship between the quantity of money in the economy and the price level, real GDP, and the velocity of money. If the velocity of money—the rate at which currency and other forms of money turn over each year—is constant, then an increase in the money supply causes an increase in the price level or an increase in real GDP, or both. The impact of the money supply on real GDP occurs more rapidly than the impact on prices. The equation has been used to explain the importance of money in the economy by many economists over the years, including Milton Friedman (who put the equation on his California license plates).

What is meant by *strict monetarist*?

A *strict monetarist* usually refers to someone with the belief that the quantity theory of money works very well in practice because the velocity of money is constant. Thus, to control the inflation rate—the increase in price levels over time—a monetarist would argue that the central bank merely needs to control the growth of the money supply: a low rate of money growth leads to a low inflation rate, and a high rate of money growth leads to high inflation rate. A strict monetarist would also argue that changes in the money supply are the dominant factor in determining inflation and causing recessions and booms.

What is the Taylor rule?

The Taylor rule is a simple and straightforward recommendation of how the Fed should conduct monetary policy. It says that the Fed should raise the interest rate when inflation increases and lower the interest rate when GDP declines. The rule also says by how much the interest rate should change: the desired rate is one-and-a-half times the inflation rate, plus one-half times the gap between GDP and its potential, plus one. But the rule was not meant to be used mechanically. Rather, it is a guide embodying important principles, more like a strategy than a mechanical formula. The rule was derived from economic theory, but over time people began to observe that it was accurate at predicting interest rates during good economic times such as the 1980s and 1990s. The economy did not do so well when the Fed deviated from the rule, as in the Great Inflation of the 1970s and the period leading up to the Great Recession of 2007–09. Thousands of articles and papers have been written on the Taylor rule. There are other monetary policy rules such as Milton Friedman's constant growth rate rule to hold money growth constant.

NOTES

1. White House, *Economic Report of the President*, Council of Economic Advisers (1962).

2. George P. Shultz and Robert Aliber, *Guidelines, Informal Controls, and the Market Place: Policy Choices in a Full Employment Economy* (Chicago: University of Chicago Press, 1966).

3. Shultz and Aliber, *Guidelines*, 19.

4. Shultz and Aliber, *Guidelines*, 25.

5. Shultz and Aliber, *Guidelines*, 31–32.

6. Shultz and Aliber, *Guidelines*, 37–38.

7. Shultz and Aliber, *Guidelines*, 39.

8. Shultz and Aliber, *Guidelines*, 41–42.

9. Friedman is referring to "Business's Plans for New Plants and Equipment, 1962–1965," *15th Annual McGraw-Hill Survey* (New York: 1962).

10. Shultz and Aliber, *Guidelines*, 42.

11. Shultz and Aliber, *Guidelines*, 43–44.

12. Shultz and Aliber, *Guidelines*, 44.

13. Shultz and Aliber, *Guidelines*, 45.

14. Shultz and Aliber, *Guidelines*, 46.

15. Shultz and Aliber, *Guidelines*, 47.

16. Shultz and Aliber, *Guidelines*, 49.

17. Shultz and Aliber, *Guidelines*, 51.

18. Shultz and Aliber, *Guidelines*, 51.

19. Shultz and Aliber, *Guidelines*, 52.

20. Shultz and Aliber, *Guidelines*, 65.

21. Friedman is referring to Paul A. Samuelson and Robert M. Solow, "Analytical Aspects of Anti-Inflation Policy," *American Economic Review* 50, no. 2 (1960): 177–94.

22. N. Gregory Mankiw, Robert M. Solow, and John B. Taylor, "Fifty Years of the Phillips Curve: A Dialog on What We Have Learned," in *Understanding Inflation and the Implications for Monetary Policy*, ed. Jeff Fuhrer, Yolanda K. Kodrzycki, Jane Sneddon Little, and Giovanni P. Olivei (Cambridge, MA: MIT Press, 2009), 77.

23. George P. Shultz, "Prescription for Economic Policy: 'Steady As You Go'" (speech, Meeting of the Economic Club of Chicago, Chicago, IL, April 22, 1971).

24. Milton Friedman, "Why the Freeze Is a Mistake," *Newsweek*, August 30, 1971.

25. "Holton Calls Nixon Move Bold," *Associated Press*, August 17, 1971.

26. "Holton Calls Nixon Move Bold," *Associated Press*.

27. "Holton Calls Nixon Move Bold," *Associated Press*.

28. Walter Stovall, "Banks, Firms Laud Nixon Move," *Associated Press*, August 17, 1971.

29. Stovall, "Banks, Firms Laud Nixon Move."

30. Willard Marriott to secretary of the Treasury George Shultz, January 15, 1973.

31. Secretary of Commerce Frederick Dent to President Richard Nixon, November 2, 1973.

32. William Safire, "Nixon on Bush," *New York Times*, July 7, 2003.

33. Gerald Ford, "Address to a Joint Session of Congress on the Economy," Washington, DC, October 8, 1974.

34. White House, *Economic Report of the President*, Council of Economic Advisers (1977).

35. Milton Friedman (speech, Meeting of the O'Hare International Bank Executives' Club, Chicago, IL, March 14, 1974).

36. Milton Friedman, "Will the Kettle Explode?" *Newsweek*, October 18, 1971.

37. John B. Taylor, "An Interview with Milton Friedman," *Macroeconomic Dynamics* 5, no. 1 (February 2001): 101–31.

38. White House, *Economic Report of the President*, Council of Economic Advisers (1990): 84.

REFERENCES
FOR FURTHER READING

Friedman, Milton, "President's Address: The Role of Monetary Policy" (lecture, 80th Annual Meeting of the American Economic Association, Washington, DC, December 29, 1967).

Mankiw, N. Gregory, Robert M. Solow, and John B. Taylor, "Fifty Years of the Phillips Curve: A Dialog on What We Have Learned," in *Understanding Inflation and the Implications for Monetary Policy*, ed. Jeff Fuhrer, Yolanda K. Kodrzycki, Jane Sneddon Little, and Giovanni P. Olivei (Cambridge, MA: MIT Press, 2009), 71–98.

Meltzer, Allan, "Under Controls, Camp David and Beyond," chap. 6 in *A History of the Federal Reserve*, vol. 2, book 2 (Chicago: University of Chicago Press, 2010).

Shultz, George P. and Robert Aliber, *Guidelines, Informal Controls, and the Market Place: Policy Choices in a Full Employment Economy* (Chicago: University of Chicago Press, 1966).

Taylor, John B., "An Interview with Milton Friedman," *Macroeconomic Dynamics* 5, no. 1 (February 2001): 101–31.

———, *First Principles: Five Keys to Restoring America's Prosperity* (New York: W. W. Norton, 2012).

White House, *Economic Report of the President*, Council of Economic Advisers (1962, 1977, and 1990).

ACKNOWLEDGMENTS

We would like to acknowledge the contributions of many members of our Hoover Institution "family" in the creation of this book. Thank you to Susan Southworth, Judy Leep, and Diana Sykes for their help in unearthing, assembling, and digitizing original materials from the Hoover Archives; to Marie-Christine Slakey and David Fedor in shepherding the manuscript into the capable hands of Danica Hodge and Alison Law at the Hoover Institution Press; to Michael Bordo, who originally prompted George Shultz to recall this era for a speech to a meeting of the Economic History Association in San Jose ("Dreams Can Be Nightmares," September 2017); to Preston Butcher for his generous support of this book and its argument that good economics leads to good policy and to good results; and to the Friedman family for letting us spend just a bit more time with our friend, and their father, Milton Friedman, in writing this.

ABOUT THE AUTHORS

George Pratt Shultz is the Thomas W. and Susan B. Ford Distinguished Fellow at the Hoover Institution. He has had a distinguished career in government, in academia, and in the world of business. He is one of two individuals who have held four different federal cabinet posts; he has taught at three of this country's great universities; and for eight years he was president of a major engineering and construction company. He attended Princeton University, graduating with a BA in economics, whereupon he enlisted in the US Marine Corps, serving through 1945. He later earned a PhD in industrial economics from the Massachusetts Institute of Technology and served on President Eisenhower's Council of Economic Advisers. From 1962 to 1969, Shultz was dean of the University of Chicago Booth School of Business before returning to Washington to serve as secretary of labor, as director of the Office of Management and Budget, and as secretary of the Treasury in the cabinet of President Nixon. Shultz was sworn in July 16, 1982, as the sixtieth US secretary of state and served until January 20, 1989. In 1989, Shultz was awarded the Medal of Freedom, the nation's highest civilian honor. He was editor of *Blueprint for America* (2016), coeditor of *Beyond Disruption: Technology's Challenge to Governance* (2018), and most recently author of *Thinking about the Future* (2019).

John B. Taylor is the George P. Shultz Senior Fellow in Economics at the Hoover Institution and the Mary and Robert Raymond Professor of Economics at Stanford University. He chairs the Hoover Working Group on Economic Policy and is director of Stanford's Introductory Economics Center. Taylor's fields of expertise are monetary policy, fiscal policy, and international economics, subjects about which he has widely authored both policy and academic texts. Taylor served as senior economist on the Council of Economic Advisers for both President Ford and President Carter, as a member of President George H. W. Bush's Council of Economic Advisers, as under secretary of the Treasury in the George W. Bush administration, and as a senior economic adviser to numerous presidential campaigns. He was a member of the Congressional Budget Office's Panel of Economic Advisers and the Eminent Persons Group on Global Financial Governance. Taylor received a BA in economics summa cum laude from Princeton University in 1968 and a PhD in economics from Stanford in 1973. His book *Global Financial Warriors* (2008) chronicles his policy innovations at the US Treasury. He received the Hayek Prize for his book *First Principles* (2012), and his recent research and writing is on international economics, including *Reform of the International Monetary System: Why and How?* (2019).

Milton Friedman (1912–2006), recipient of the 1976 Nobel Memorial Prize for economic science, was a senior research fellow at the Hoover Institution from 1977 to 2006. He was also the Paul Snowden Russell Distinguished Service Professor Emeritus of Economics at the University of Chicago, where he taught from 1946 to 1976, and a member of the research staff of the National Bureau of Economic Research from 1937 to 1981.

He is widely regarded as the leader of the Chicago School of economics. Friedman wrote extensively on public policy, always with a primary emphasis on the preservation and extension of individual freedom. His most important books in this field are *Capitalism and Freedom* (1962); *Bright Promises, Dismal Performance* (1983); *Free to Choose* (1980, with Rose Friedman); and *Tyranny of the Status Quo* (1984, with Rose Friedman). Friedman was active in public affairs, serving as an informal economic adviser to Senator Barry Goldwater in his unsuccessful presidential campaign in 1964; to Richard Nixon in his successful 1968 campaign and during his presidency; and to Ronald Reagan during and after his winning 1980 campaign, subsequently serving as a member of the President's Economic Policy Advisory Board. He was awarded the Presidential Medal of Freedom in 1988 and received the National Medal of Science the same year. Friedman passed away on November 16, 2006.

More than 1,500 of Friedman's articles and speeches are available through the Hoover Institution Library & Archives, online at miltonfriedman.hoover.org, and excerpted in *Milton Friedman on Freedom* (Hoover Institution Press, 2017).

INDEX

A. H. Robins, 24

Ackley, Gardner, 2

Aliber, Robert, 1, 15, 65

Allied powers, 100

American Economic
    Association, 15

American Recovery and
    Reinvestment Act, 71

bailouts
    of Penn Central, 17
    of Wall Street, 62

Banking, Housing, and Urban
    Affairs Committee, 66

Belgrade, 50

Black (Justice), 6

Boskin, Michael, 58f

Bretton Woods Agreement, 35
    history of, 99–100

Brezhnev, Leonid, 30–31

Budget Act of 1974, 86

budgets
    cuts in, 84
    in economic strategy,
        84–86
    off-budget financing, 85

Burns, Arthur, 15, 23f, 31,
    46–48, 65
    on GDP, 48–49
    on inflation, 20, 73–74,
        77–78
    Nixon and, 48, 73–79
    on Penn Central, 17–18
    Shultz on, 19f
    on trade unionism, 74–75
    on wage and price controls,
        20–21, 66
    on welfare programs, 75

Bush, George H. W., 70
    Council of Economic
        Advisers and, 58f
    tax code reforms under,
        57–58
Bush, George W.
    on Keynesianism, 60
    tax cuts under, 60
    Taylor and, 61f
business cycle, 20–21
Byrd, Harry, 24

Cabinet Room, 52f
Camp David, 23f, 35
Carter, Jimmy, 44
    on consumer credit, 50
    on inflation, 69
centralization, 12–13
Chase Manhattan, 39
Cheney, Dick, 27, 55
classical economic policy, 99
Clinton, Bill
    on government, 70
    stimulus under, 59
compliance measures, 5
Connally, John, 23f, 25,
        55, 66
constant growth rate
    rule, 102

consumer credit
    Carter on, 50
    controls on, 50
consumer price index (CPI),
        25, 42, 44, 46
controls. See wage and price
        controls
Coordinating Committee
        on Economic Policy,
        56–57, 69
    action of, 91
    on energy, 88–89
    on monetary policy,
        89–90
    on oil price controls, 88
    of Reagan, 81–93
    recommendations of,
        85–86
    on regulations, 45, 87
    on regulatory reforms,
        87–88
    Shultz at, 45
    signatures of, 94f
    steady course maintained
        by, 91–92
    on tax policy, 85–86
Cost of Living Council, 25, 67
Council of Economic Advisers,
        6, 23f, 52f

Bush, G. H. W., and, 58f
  public opinion and, 13
Council on Wage and Price
    Stability, 69, 88
  elimination of, 45
  passage of, 68
CPI. *See* consumer price
  index

deficit spending, 100
deflation, 61–62
Dent, Frederick, on wage and
  price controls, 29
Department of Defense, 17
Department of Energy, 89
depreciation, 86–87
Diocletian, 95
disinflation, 51f
Dole, Bob, 59
dollar, US
  credibility of, 90
  overvaluation of, 35
  price controls of, 37–38
Dunlop, John, 2, 55, 67, 68

Economic Club of Chicago, 66
Economic Growth and Tax
  Relief Reconciliation
  Act of 2001, 70

Economic Policy Board, 91
Economic Recovery Tax Act
  of 1981, 69
*Economic Report of the Presi-
  dent,* 12, 33–34, 95–96
Economic Stabilization Act,
  22–23, 68
  passage of, 65
Economic Stimulus Act of
  2008, 70–71
economic strategy
  budgets in, 84–86
  coherence in, 83
  consistency in, 83
  guiding principles of,
    82–84
  initiative in, 84
  long-term, 84
  problems faced by, 83
  public opinion and, 84
  of Reagan, 81–93
  steady course in, 91–92
  tax policy and, 86–87
Eisenhower, Dwight, 17
employment, full, 8. *See also*
  unemployment
energy, Coordinating Com-
  mittee on Economic
  Policy on, 88–89

Energy Policy and
    Conservation Act,
    68–69
Euro-dollar market, 41
Executive Branch, 91

Federal Energy Administration,
    68–69
Federal Energy Office, 42
federal funds rate, 45–46
    Taylor rule and, 51f
Federal Regulatory
    Agency, 46f
Federal Reserve, 20, 45–46
    Friedman on, 40
    Greenspan at, 59
    independence of, 90
    on interest rates, 59, 61
    monetary policy of, 62
    transparency at, 59
    Volcker at, 55–56
First Principles: Five Keys
    to Restoring America's
    Prosperity (Taylor), 62
fixed exchange rates, 37, 39
    breakdown of, 100
    Rockefeller on, 40–41
floating exchange rates,
    37, 100

Ford, Gerald, 31, 32f, 68, 98–99
    recession during presidency
    of, 44
Francis, Darryl, 47
free market
    gold window and, 37–38
    wage and price guideposts
    and, 11–12
Friedman, Milton, 7f, 32,
    36f, 96
    on Federal Reserve, 40
    on Great Inflation, 49–50
    on inflation, 2–7, 15–16
    on international economic
        policy, 39–40
    on Phillips curve, 97
    on Reagan, 49–50, 53
    on recessions, 43–44
    Shultz and, 37–40
    Solow and, 13–15
    Taylor and, 47–53, 57–58
    on Volcker, 49–50
    on voluntary controls, 5–6
    on wage and price
        controls, 41
    on wage and price
        guideposts, 3–7
    on wage-price freeze, 38
full employment, 8

GDP. *See* gross domestic product

The General Theory (Keynes), 99

Gingrich, Newt, 59

global financial crisis of 2007–09, 62

gold window, 27
  free market and, 37–38
  Nixon on, 35–38, 100

Great Inflation, 47, 102
  end of, 49–50
  Friedman on, 49–50

Great Recession, 102

Great Society, 1

Greenspan, Alan, 33
  at Federal Reserve, 59
  on price stability, 59

gross domestic product (GDP)
  Burns on, 48–49
  growth of, 34f

guidelines. *See* wage and price guideposts

Guidelines, Informal Controls, and the Market Place, 2, 65

guideposts. *See* wage and price guideposts

Harlow, Bryce, 18

Hartley, Fred, 41

Hicks, John, 99

Holton, Linwood, 23

Hoover Archives, 20

Hoover Institution, 57

housing market, 60
  bubble, 62

Hubbard, Glenn, 60

incomes policy, defining, 95–96

inflation, 1. *See also* Great Inflation
  Burns on, 20, 73–74, 77–78
  Carter on, 69
  deceleration of, 69
  decision-making and, 76
  Friedman on, 2–7, 15–16
  market power and, 10
  monetarism and, 101
  as monetary phenomenon, 2–3
  monetary policy and, 59
  open, 3–7
  persistence of, 76
  Phillips curve and, 97
  premature, 11
  under Reagan, 50
  recession and, 47

inflation (*continued*)
   Solow on, 8
   suppressed, 3–7
   unemployment and, 9,
      74, 78
   wage and price controls
      and, 18–19
   wage-price freeze and, 38
interest rates, 41, 51
   Federal Reserve on, 59, 61
   Taylor rule and, 51f
international economic
      policy, Friedman on,
      39–40
International Monetary Fund,
      establishment of, 100
interventionism, 62, 63
Iranian Revolution, 45
Iran-Iraq War, 89

Jobs and Growth Tax Relief
      Reconciliation Act, 70
Johnson, Lyndon, 1, 41–42
Joint Economic Committee,
      39, 65–66, 100

Kemp, Jack, 52f
Kemp-Roth cuts, 86
Kennedy, John F., 1, 41–42

Keynesianism, 48
   Bush, G. W., on, 60
   defining, 99

Laffer, Arthur, 52f
Laidler, David, 36f
Landrum-Griffin Act, Burns
   on, 75
leadership, 92–93
Lynn, James T., 52f

marginal tax rates
   under Reagan, 56
   reduction of, 56
market power
   defining, 97–98
   inflation and, 10
   Solow on, 10, 98
Marriott, J. Willard, on wage
   and price controls,
   27–29
McCracken, Paul, 23f, 66
McGraw-Hill survey, 9
Meltzer, Allan, 2
Miller, G. William, 49
monetarism, 47–48
   inflation and, 101
   strict, 101
monetary growth, 89–90

monetary policy
 Coordinating Committee
  on Economic Policy on,
  89–90
 of Federal Reserve, 62
 inflation and, 59
 in 2000s, 61

NAFTA (North American Free
 Trade Agreement), 59
National Bureau of Economic
 Research, 20
National Security Council, 91
Natural Gas Policy Act of
 1978, 89
New Economic Program, 35,
 66–67
Nixon, Richard, 23f
 Brezhnev and, 30–31
 Burns and, 48, 73–79
 on gold window, 35–38, 100
 New Economic Policy
  of, 66
 Shultz and, 30–31
 on wage and price controls,
  18–19, 30, 33f, 42, 55
 wage-price freeze of, 18–19
North American Free Trade
 Agreement (NAFTA), 59

Obama, Barack, 62, 71
off-budget financing, 85
Office of Management and
 Budget, 66, 85
oil price controls, 40, 69
 Coordinating Committee on
  Economic Policy on, 88
Omnibus Budget Reconcilia-
 tion Act, 70
open inflation, 3–7

Packard, David, 17
Pay Board, 24, 96
 abolition of, 30
Penn Central
 bailout of, 17
 Burns on, 17–18
 Shultz on, 18
Pentagon, 17
Peterson, Peter, 66
Phillips, A. W., 97
Phillips curve, 15, 48
 defining, 97
 Friedman on, 97
 inflation and, 97
 unemployment and, 97
premature inflation, 11
Price Commission, 24, 96
 abolition of, 30

price controls. *See* wage and
   price controls
price guideposts. *See* wage
   and price guideposts
price setting, 1
price stability, Greenspan
   on, 59
public opinion, 12
   Council of Economic
      Advisers and, 13
   economic strategy and, 84
   of Reagan, 50
public sector, trade unionism
   in, 74–75

Quadriad, 73
quantitative easing, 62
quantity theory of
   money, 101

rationing, 42–43
Reagan, Ronald, 44f, 64
   Coordinating Committee
      on Economic Policy of,
      81–93
   economic strategy of, 81–93
   Friedman on, 49–50, 53
   inflation under, 50
   marginal tax rates under, 56

public opinion ratings
   of, 50
recession during presidency
   of, 44–45
Shultz and, 69, 81–93
tax cuts under, 69–70
Volcker on, 46
recessions
   Friedman on, 43–44
   Great Recession, 102
   inflation and, 47
   Reagan and, 44–45
Regulation Q, 41–42
regulatory reforms, 87–88
Reynolds, Richard, Jr., 24
Reynolds Metal Company, 24
Rivlin, Alice, 59
Robins, Claiborne, 24
Rockefeller, David, 39
   on fixed exchange rates,
      40–41
Rumsfeld, Donald, 27, 31,
   55, 67

Safire, Bill, 31, 66
Samuelson, Paul A., 15
saving, 21
Schmalensee, Richard, 58f
Schmidt, Helmut, 17

Senate Banking Committee, 68

Shultz, George, 1, 15, 23f, 27, 36f, 52f
 Burns and, 19f
 at Coordinating Committee on Economic Policy, 45
 Friedman and, 37–40
 Nixon and, 30–31
 on Penn Central, 18
 Reagan and, 69, 81–93
 "Steady As You Go" speech of, 17, 66
 on wage and price controls, 65

Simon, Bill, 42–43

Smithsonian Agreement, 27

Social Security, 60

Solow, Robert, 2, 14f
 Friedman and, 13–15
 on guideposts, 8–14
 on inflation, 8
 on market power, 10, 98
 on unemployment, 15
 on wage and price guideposts, 8–14, 16

Soviet Union, 30–31

Spong, William, Jr., 22

Stanford, 57

"Steady As You Go" speech (Shultz), 18, 66

Stein, Herb, 30, 66

Stigler, George, 2

stimulus, 62, 70
 under Clinton, 59

stock market, wage-price freeze and, 22

strict monetarists, 101

suppressed inflation, 3–7

Synthetic Fuels Corporation, 89

tax code reforms, 45
 under Bush, G. H. W., 57–58

tax cuts
 under Bush, G. W., 60
 under Reagan, 69–70
 Taylor on, 62

tax policy
 Coordinating Committee on Economic Policy on, 85–86
 economic strategy and, 86–87

tax rebates, 68–71

Tax Reform Act, 56, 70

tax-based incomes policies (TIPs), 96

Taylor, John, 15, 32f, 33, 36f,
    58f, 70
    Bush, G. W., and, 61f
    Friedman and, 47–53, 57–58
    on tax cuts, 62
    at Treasury, US, 60
Taylor rule, 63
    defining, 101–2
    federal funds rate and, 51f
    interest rates and, 51f
TIPs (tax-based incomes poli-
    cies), 96
trade unions
    Burns on, 74–75
    in public sector, 74–75
    on wage and price guide-
        posts, 11

unemployment, 8, 16
    inflation and, 9, 74, 78
    Phillips curve and, 97
    rates of, 49f
    Solow on, 15
Union Oil Company, 41
unions. *See* trade unions

Vietnam War, 1
V-loans, 17
Volcker, Paul, 39, 45–46, 50, 66

    at Federal Reserve, 55–56
    Friedman on, 49–50
    on Reagan, 46
voluntary controls, Friedman
    on, 5–6

wage and price controls
    Burns on, 20–21, 66
    defining, 95–96
    Dent on, 29
    Friedman on, 41
    inflation and, 18–19
    Marriott on, 27–29
    Nixon on, 18–19, 30, 33f,
        42, 55
    for oil, 40, 69, 88
    phase I of, 96, 100
    phase II of, 96
    phase III of, 67, 97
    phase IV of, 67, 97
    Shultz on, 65
    on US dollar, 37–38
wage and price guideposts, 65
    causality and, 16
    defining, 95–96
    free market and, 11–12
    Friedman on, 3–7
    goals of, 10
    Solow on, 8–14, 16

unions on, 11
weaknesses in, 12–13
wage setting, 1
wage-price freeze
    Friedman on, 38
    inflation and, 38
    of Nixon, 18–19
    popularity of, 22
    stock market and, 22
    success of, 24–25
Wall Street, bailout of, 62

Wall Street Journal, 22
Weidenbaum, Murray, 52f
Weinberger, Caspar, 66
welfare programs, 59
    Burns on, 75
Whip Inflation Now campaign, 31–32, 68
    history of, 98–99
World Trade Organization (WTO), 59
World War II, 100